IR Part-66

Aircraft Maintenance Licence

Distance Learning Modules

Module 8 – Basic Aerodynamics

Chris Strike

Copyright © Cardiff and Vale College 2011
Issue: 1-12

Cardiff and Vale College

Published by:
Cardiff and Vale College
International Centre for Aerospace Training
Cardiff Airport Business Park
Port Road
Rhoose
Vale of Glamorgan
CF62 3DP
Wales
United Kingdom

+44 (0) 1446 719821

info@part66.co.uk

www.part66.com

Copyright © Cardiff and Vale College 2011

All rights reserved. No part of this publication may be reproduced or transmitted in any form or by any means, electronically or mechanically, including photocopying, recording or any information storage or retrieval system, without permission in writing from the publisher or a license permitting restricted copying.

In the United Kingdom such licenses are issued by the Copyright Licensing Agency:

90 Tottenham Court Road, London W1P 4LP.

Liability

Whilst the advice and information in this book are believed to be true and accurate at the date of going to press, neither the author nor the publisher can accept any legal responsibility or liability for any errors or omissions that may be made.

Author: Chris Strike

IR Part-66   Module 8

Basic Aerodynamics

Knowledge Levels

The basic knowledge requirements for categories A, B1 and B2 certifying staff are indicated in the contents list of the module notes by knowledge level indicators 1 or 2 against each chapter.

Level 1

A familiarisation with the principal elements of the subject

Objectives

The student should be familiar with the basic elements of the subject.
The student should be able to give a simple description of the whole subject using common words and examples.
The student should be able to use typical terms.

Level 2

A general knowledge of the theoretical and practical aspects of the subject. An ability to apply that knowledge.

Objectives

The student should be able to understand the theoretical fundamentals of the subject.
The student should be able to give a general description of the subject using, as appropriate, typical examples.
The student should be able to use mathematical formulae in conjunction with physical laws describing the subject.
The student should be able to read and understand sketches, drawings and schematics describing the subject.
The student should be able to apply knowledge in a practical manner using detailed procedures.

IR Part-66   Module 8

Basic Aerodynamics

Contents

Chapter One – Physics of the Atmosphere

Knowledge Level: A - 1, B1 - 2, B2 - 2

The Atmosphere	1
Air Density	2
Atmospheric Pressure	3
Measurement of Pressure	3
Air Temperature	4
Measurement of Air Temperature	4
Effects of Air Pressure on Temperature and Density	5
Effect of Humidity	6
The Layers of the Atmosphere	6
International Standard Atmosphere	9
Q Codes	12
Pressure Altitude	13
Airspeed	14
Revision	16

Chapter Two – Aerodynamics

Knowledge Level: A - 1, B1 - 2, B2 - 2

Airflow	23
Airflow Around a Body	23
Boundary Layer	25
Laminar & Turbulent Flow	26
Boundary Layer Separation	27
Free Steam Flow	28
The Aerofoil	28
Upwash	30
Downwash	31
Vortices	33
Stagnation	35
Configuration & Aerodynamic Performance	35
Definition of Terms	35
Camber	36
Chord	37
Thickness/Chord Ratio	38
Fineness Ratio	38
Mean Aerodynamic Chord	39
Angle of Attack	40
Angle of Incidence	41
Wash In	42
Wash Out	42
Centre of Pressure	42
Aerodynamic Centre	43
Wing Shape & Aspect Ratio	44
Wing Shape	46
Taper Ratio	46
Straight Rectangular	46
Elliptical	47
Tapered	47
Wing Sweep Angle	48
Swept Back Wing	48
Crescent Wing	49
Forward Swept Wing	50
Delta Wing	51
Slender Delta	51
Polymorphic Wing	52

Drag	53
Profile Drag	53
Form Drag	53
Skin Friction	54
Interference Drag	54
Calculation of Profile Drag	54
Induced Drag	55
Calculation of Induced Drag	57
Total Drag	58
Thrust & Weight	58
Thrust	58
Weight	59
Aerodynamic Result	61
Generation of Lift & Drag	62
Lift	62
Coefficient of Lift	64
Drag	65
Total Drag	67
Coefficients of List & Drag	68
Polar Curve	69
Stall	69
The Effects of Wing Sweep	71
The Effect of Sweep Back on Lift	72
The Effect of Sweep Back on Drag	73
The Effect of Sweep Back on Stalling	73
High Speed Stall	75
Pitch Up	76
Pitch Down	77
Aerofoil Contamination	77
Ice	77
Super-Cooled Clouds	77
Affects of Ice Accretion	78
Revision	81

Chapter Three – Theory of Flight

Knowledge Level: A - 1, B1 - 2, B2 - 2

Relationship between Lift, Weight, Thrust & Drag	91
Lift	91
Weight	92
Thrust	92
Drag	92
Arrangement of the Four Flight Forces	93
Relationship Between the Four Forces	95
No Power Gliding	96
Effect of Wind	99
Effect of Weight	100
Glide Ratio	100
Steady State Flights - Performance	101
Straight & Level Flight	101
Climbing	102
Powered Descent	103
The Sideslip	105
Theory of the Turn	105
Centripetal & Centrifugal Forces	105
Aircraft Turning & Banking	108
The Level Turn	109
Climbing Turn	112
Gliding Turn	113
Descending Turn	113
Influence of Load Factor	114
Stall	117
Effect of Altitude on Stalling Speed	118
Effect of Weight on Stalling Speed	118
Effect of Flaps & Slats on Stalling Angle & Speed	118
Effect of Load Factor on Stalling Speed	118
Effect of Engine Power on Stalling Speed	119
Effect of Icing on Stalling Speed	119
Effect of Centre of Gravity Position on Stalling Speed	120
Effect of Aileron Use	120
Wing Tip Stall	121
Stall Strips	121
Stall Sensing	121
Deep Stall	122
High Speed Stall	122

Lift Augmentation		123
Trailing Edge Flaps		123
The Plain Flap		124
The Split Flap		124
The Zap Flap		125
The Fowler Flap		126
The Slotted Flap		126
The Slotted Fowler Flap		127
Pitching Moment		129
Leading Edge Devices		129
Slots		130
Slats		130
Fixed Slats		131
Moveable Slats		131
Leading Edge Flaps		133
Leading Edge Droop		133
Flaps & Slats		134
The Blown Flap		135
The Jet Flap		135
Use of Flaps for Take-off		135
Use of Flaps in the Air		135
Spoilers		136
Vertical Lift Control		137
Active Load Alleviation		137
Secondary Flight Controls		137
Revision		139

Chapter Four – Flight Stability & Dynamics

Knowledge Level: A - 1, B1 - 2, B2 - 2

Flight Stability & Dynamics	149
Static Stability	149
Longitudinal Stability	152
Longitudinal Control	155
Neutral Point & the Static Margin	157
Manoeuvre Point & the Manoeuvre Margin	157
Trim Drag	157
Longitudinal Dihedral	158
Stick-fixed Stability	158
Canard Configuration	158
High Speed Aircraft	160
Lateral Stability	160
Roll Damping	160
Lateral Dihedral	161
Stability During a Sideslip	162
Effect of Sweepback on Lateral Stability	162
Anhedral	163
Low Wing Aircraft	164
High Wing Aircraft	164
Keel Surface	165
Shielding Effect of Fuselage	165
Effect of Flap Deployment on Lateral Stability	165
Lateral Control	166
Adverse Yaw	167
Directional Stability	168
Weathercock Effect	168
Effect of Keel Surface	168
Longitudinal Centre of Pressure	169
Design of the Fin	169
Centre of Gravity Position	169
Shielding Effect	170
High Speed Aircraft	170
Directional Control	170
Cross Links between Axes of Control	170
Stick Free Stability	171
Affect of Altitude	171
Dynamic Stability	172
Longitudinal Dynamic Stability	175
Effect of Altitude	176
Phugoid (Porpoising)	176
Lateral Dynamic Stability	177
Dutch Roll	178

Spiral Stability	180
The Spin	181
Directional Dynamic Stability	182
Stability of Propeller Driven Aircraft	182
Stability Design Features	183
Unstable Aircraft	184
Basic Autopilot	185
Revision	187
Glossary	197

Physics of the Atmosphere

The Atmosphere

The Earth's atmosphere is an envelope of gas held in place by gravity that, together with the magnetosphere, shields us from the worst effects of cosmic and solar radiation. Over time, it has been modified to its present composition by out-gassing from volcanoes, rocks, and meteorites and from various living organisms. Below 500km (310miles) the gas composition by percentage volume is approximately 78% nitrogen, 21% oxygen with the remaining 1% made up from traces of argon, carbon dioxide, hydrogen, neon, helium, methane, xenon, ozone, radon and krypton. The lower atmosphere also holds water vapour that can vary in content between 0 and 4% by volume.

The atmosphere can be considered as consisting of five theoretical layers. Ascending from the Earth's surface these are, the *Troposphere*, the *Stratosphere*, the *Mesosphere*, the *Thermosphere* and finally the *Exosphere* where the atmosphere gradually merges into the extremely rarefied regions of space. Most of the private and commercial aircraft flights occur in the region of the troposphere. High speed military aircraft and until recently, the Concorde, often operate at altitudes within the stratosphere. It is difficult to quote an exact figure for the total height of the atmosphere, as it is still just detectable several thousands of kilometres above the Earth's surface. The exact altitudes occupied by the various layers are again the subject of debate by various authorities and are subject to geographical and seasonal changes. For the purposes of this syllabus it is the regions occupied by the troposphere and the stratosphere that will be of the most interest to us. I will include a brief description later of the regions above this merely to paint you a fuller picture.

CHAPTER ONE
PHYSICS OF THE ATMOSPHERE

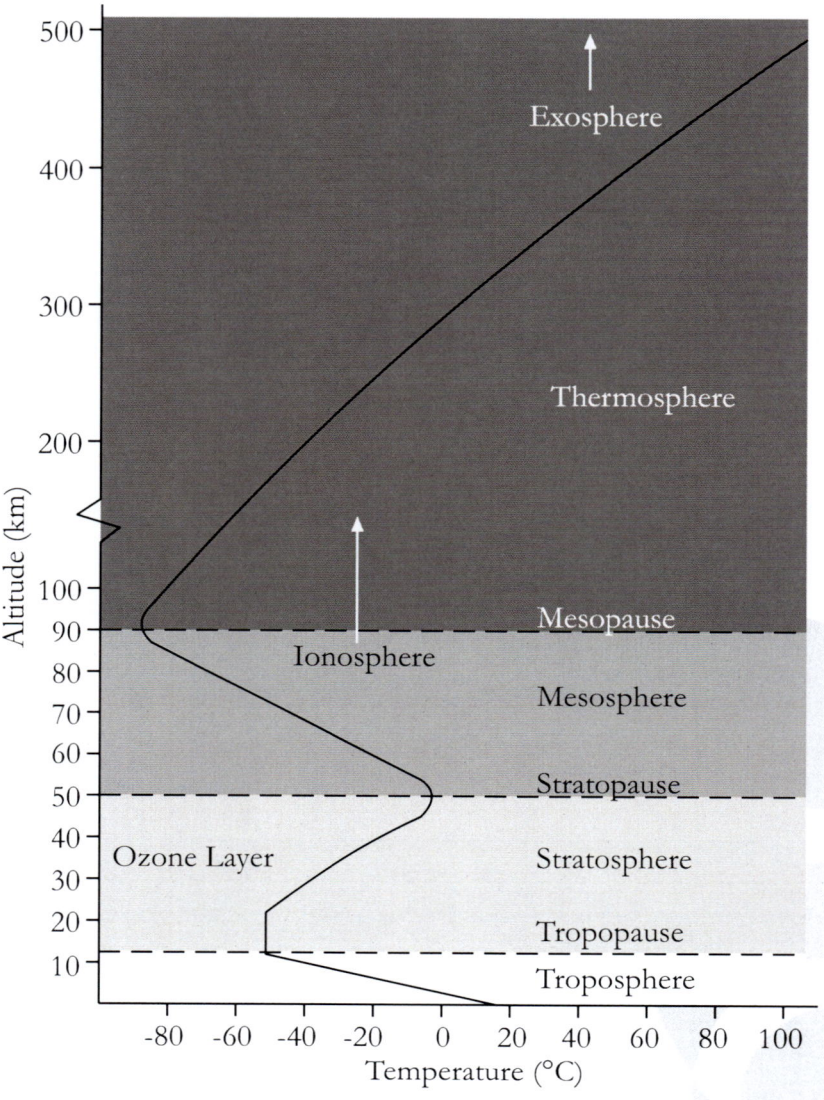

Figure 1.1 - The Atmosphere

Air Density

Before we enter into a description of each layer of the atmosphere it would be useful to consider just why air pressure, temperature and density changes occur in them. Air is a gas and is thus compressible. If you consider an imaginary cardboard box having a volume of one cubic metre and we were to fill it with air molecules then it would not be unreasonable to expect that if you then filled it with compressed air you would have more air molecules in the box than before. The air molecule does have mass so the more of them you have in the box the greater the mass of air it would contain. At sea level the mass of air in our box would be 1.225kg. Density is mass divided by volume so the density of our air at sea level is 1.225kg/m^3.

Atmospheric Pressure

You may be surprised at the above and wonder why air has such a mass when you cannot even see it. The reason is that at sea level it is compressed under a pressure of about 100kPa, often referred to as one bar or one Atmosphere. We can also express this pressure in millibars, a millibar being one thousandth of a bar (0.1kPa). Expressed in imperial measurement this would be around 14.7lb/in². The question is, why the pressure?

Pressure, when applied to a fluid such as air is defined as force per unit area. Due to the free flowing nature of fluids, the force exerted by a fluid is always at right angles to any surface in contact with it. The force exerted on a unit area of any such surface is defined as the fluid's pressure.

$$\text{Pressure} = \frac{\text{Force}}{\text{Area}}$$

Remove the sides of our imaginary cubic metre box and now imagine the cubic metre of air to be a cube of foam rubber. We now stack further similar volume cubes one on top of the other until we reach the top of the known atmosphere. Can you see that the cube at the bottom of the column is supporting all the cubes above it and will thus be squeezed by the gravitational force of them, it will be under pressure. The base of the cube is one square metre so if you divide that into the force you will have the pressure. If you now look at a cube halfway up the column, it is not supporting the weight of quite so many cubes, half in fact. So, it will not be squeezed quite to the extent as the cube at the bottom. At the top of the column the top cube is supporting nothing at all and will only experience its own gravitational pull, not squeezed at all. That should explain why we have atmospheric pressure and why it decreases as we gain altitude.

If you now imagine that we adjust the dimensions of the squeezed cube at the bottom back to one cubic metre without altering the pressure in it there will be more foam rubber in it than in the cubic metre at the top of the column. Hence the density of the column reduces with increasing altitude.

Measurement of Pressure

The pressure gauges encountered in everyday use only measure pressures above atmospheric pressure. A tyre pressure gauge for example may read zero when lying on a bench but the workshop it is in may have an atmospheric pressure of 15psi. This pressure does not register on the gauge. This type of gauge reads what is called *Gauge Pressure*.

For example, if we now check the air pressure in a tyre and observe that the gauge reads 30psi it is reading the gauge pressure only and that is 30psi higher than atmospheric pressure. If you were to add the prevailing atmospheric pressure of 15psi to the gauge reading you would conclude that the actual pressure should be 45psi. This figure is referred to as the *Absolute Pressure*.

As far as the tyre is concerned we would only be interested in the gauge pressure because that gives us an indication of the difference between the tyre's internal pressure and the pressure already present in the workshop. So, we need to be clear that the figures we quote for atmospheric pressures are absolute pressure figures. In the example given above, the figure of 15psi quoted for atmospheric pressure is the pressure above the zero pressure of a perfect vacuum. A useful formula for you to remember is:

> Absolute Pressure = Gauge Pressure + Atmospheric Pressure

The question remains as to how we can measure an absolute pressure in the room you are sitting in. You will have to use a **Barometer**, as this is the only instrument capable of measuring absolute pressure. Examiners like this question. You may have seen a Fortin barometer in someone's house being used to check the air pressure to indicate weather. Aircraft use an aneroid barometer to measure absolute pressure and thus altitude. There is one final point to be made. The atmospheric pressures that we record are for still or static air only. Moving air will have something called dynamic pressure as well and that is over and above the figure we are after. So, atmospheric pressures are absolute, **Static** pressures.

Air Temperature

Now imagine that the surface of the Earth is warm, it is certainly hot in the middle so the surface is warm. The bottom cube conducts heat from the surface and also absorbs some radiant heat from the Sun and as a result it warms up. It then passes some of its heat up to the cube above that in turn warms up and passes some of this heat up. On each pass, less heat is available to pass up so, the temperature of the air reduces with increasing altitude. There will come a point where so little heat is passed up that the temperature will become almost constant. This occurs at around 36,090ft (11km). As altitude continues to increase, the affect of the Sun's radiant heat will eventually begin to drive the temperature up again. At the very top of the column the full impact of the Sun's radiation will drive the temperature up to around 1,000°C.

Measurement of Air Temperature

If you recall, there were two measurements for static air pressure, absolute and gauge. We have a similar set up for temperature. If the temperature in the room you are in is 15°C this merely indicates that the room is 15°C above a nominated zero figure based on the freezing point of pure water. It is not a true temperature reading. The lowest value for heat is a point where all atomic particle motion ceases. Simply, no heat at all! This point is known as **Absolute Zero**. This relates to a Celsius scale reading of **-273.15°C**. There is obviously a need for an absolute temperature scale to deal with this and that is known as the **Kelvin** scale.

Using Kelvin, absolute zero equates to -273.15°C so the nominated point where pure water freezes in absolute terms will be 273.15K and the 15°C in our example will become 288.15K. Be aware of the temperature scale in use when describing atmospheric temperatures. You can easily convert Celsius readings to Kelvin by adding 273.15 to the Celsius reading. A final point here is that America frequently uses the Fahrenheit scale instead of Celsius so they have a different absolute temperature scale and that is the Rankine scale.

Effects of Air Pressure and Temperature on Density

We have seen how pressure can affect density but temperature can do this too. As the air gets colder it will contract. This gives more molecules per cubic metre. Density will rise.

If the air gets warmer it will expand, density will fall. When temperature is working in concert with pressure we have to be careful when considering the density change. For example, as altitude increases from sea level, the air temperature falls but so also does the air pressure. All the decreasing temperature does is reduce the rate at which the air density falls. Above 36,090ft, the air temperature remains constant so, the air density reduces at a faster rate under the sole influence of the reducing atmospheric pressure.

Gas laws govern the effects discussed above. These are: Boyle's Law, Charles' Law, the Pressure Law and the Combined gas law. These are a feature of Module 2 - Physics of course so I will only give you a very brief description of the respective relationships they reveal.

Boyle's Law: The volume of a fixed mass of gas at constant temperature is inversely proportional to its pressure. Under this condition, if pressure decreases the volume increases and as a result the density decreases.

Charles' Law: The volume of a fixed mass of gas at constant pressure is directly proportional to its absolute temperature. Under this condition if the absolute temperature decreases the volume reduces and as a result the density will increase.

Pressure Law: The pressure of a fixed mass of gas at constant volume is proportional to its absolute temperature. Under this condition if the absolute temperature decreases the pressure will decrease and density will also decrease.

Combined Gas Law: Taken in isolation the effects seem fairly clear but we have the problem that they all act together. When they do we discover something called the gas constant as follows:

$$\frac{Pressure \times Volume}{Temperature(Abs)} = A\ Constant$$

CHAPTER ONE
PHYSICS OF THE ATMOSPHERE

You can play around with this equation as long as the answer always stays the same. If you increase the pressure for example, you need to reduce the volume at constant temperature. If you increase the volume, you will need to reduce the pressure at constant temperature. Increase the absolute temperature and you need to increase either the pressure or the volume or both. Imagine the cubic metre box full of air molecules and you can start to relate the density changes that will occur.

Effect of Humidity

Air will absorb water forming a solution of air and water vapour. The higher the air temperature becomes, the more water may be absorbed until it reaches a saturation level. As the air temperature falls, excess moisture will be precipitated out. At the higher altitudes around the Tropopause the air temperature is too low to contain much moisture so the air at these altitudes will be dry. The water molecule consists of hydrogen and oxygen. Air molecules are predominantly made of nitrogen and oxygen. The nitrogen atom is seven times the mass of the hydrogen atom. Consider our imaginary cubic metre of air. If it consists of a fair proportion of water vapour there will be an increased number of hydrogen atoms and less nitrogen atoms. So, our cubic metre will have less mass and thus less density. Put another way, the density of water vapour is five eighths that of dry air so the more water vapour present in the air the less dense the air will become. The forces of lift, drag and thrust during flight reduce as air becomes more humid.

The Layers of the Atmosphere

We have been a while getting to this topic but I hope you will agree we are better off having knowledge of the few basics above. The Troposphere occupies the region from sea level up to an altitude that varies from about 19km (12 miles) at the equator to around 8km (5 miles) at the N and S poles. It is the region in which most of the familiar weather conditions we experience will occur. As you ascend from sea level up through the troposphere, the air temperature decreases steadily with height at a constant *lapse rate* of 6.5°C per km (1.98°C per thousand ft) levelling out at the top to about -80°C in equatorial regions and about -40°C in the Polar Regions in summer. The top of the troposphere is called the *Tropopause* and this marks the foot of the stratosphere. The air temperature now remains relatively constant as you pass through the lower levels of the stratosphere.

The atmospheric pressure and thus the air density fall as you ascend from sea level, although not at a constant rate. The atmospheric pressure at sea level is ideally 101.3kPa or 1013 millibars (14.7 lb/sq in) and this reduces to about 50kPa (7.35 lb/sq in) at 5.5km (18,000 ft) and 25kpa (3.5 lb/sq in) at around 10km (33,000 ft). The decreasing air temperature up to the tropopause reduces the rate at which the air density decreases with height. The density will decrease at a more rapid rate with increases in height above the tropopause because the air temperature will then be relatively constant at an average -56.5 °C up to an altitude of about 20km (12.4 miles or 65,617ft). The rate of density reduction will then increase as it is affected by the decrease in atmospheric pressure only. The loss of lift from aerofoils and the thrust from engines follows a similar rate change as both depend on air density for performance.

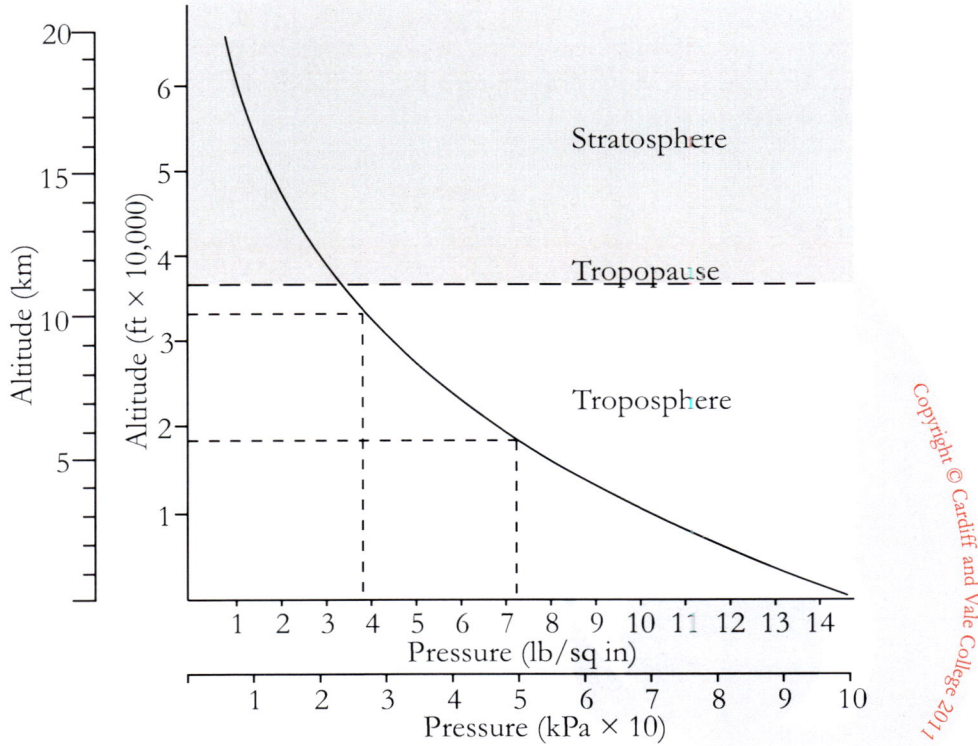

Fig 1.2 - Atmospheric Pressure

The stratosphere reaches above the tropopause to a height of about 40km (25 miles) and contains the same gas composition by volume. This region is very dry and is usually noted for its smooth flying conditions but clear air turbulence can be encountered. The **Ozone layer** may be found in this region at a height of around 30km (19 miles). As you ascend through the stratosphere the air temperature initially remains constant but will then slowly increase until it reaches about 0°C at the top of the region. This is called the **Stratopause** and is considered to be the foot of the mesosphere. Atmospheric pressure and thus air density continue to fall through the stratosphere.

The mesosphere extends up to an altitude of about 90km (56 miles). The air temperature in this region begins to decrease with height to a minimum of around -100°C at the mesopause, which is at the foot of the thermosphere. For your interest only, the air density at this altitude will be about 0.000001 kg/m³ as opposed to the air density at sea level, which is 1.225kg/m³ at 15°C. As you ascend through the thermosphere the temperature now begins to rise again reaching around 1,000°C as you pass into the exosphere and then into interplanetary space.

Before we leave our journey through the Earth's atmosphere we should make mention of the *Ionosphere*. This is a region that commences in the upper mesosphere and extends into the thermosphere. The gas in this region becomes ionised due to bombardment by solar radiation that removes outer shell electrons from its atoms leaving them with a positive electrical charge. It is because this electrified layer has the capacity to reflect radio waves that makes long distance radio communication possible. The region is affected by sunspot activity.

CHAPTER ONE
PHYSICS OF THE ATMOSPHERE

The International Standard Atmosphere (ISA)

Hopefully you will have by now noticed how approximate and variable the Earth's atmosphere can be in terms of air pressure, temperature, humidity and density at any particular time or location on the planet. This presents a problem when trying to assess the performance of machines that rely on the density of the air for their successful operation. Aircraft rely on air density to produce lift from their wings. Aircraft engines also rely on air density to produce thrust or power. Whereas an aircraft may lift off successfully at a given speed and thrust from a 10,000ft runway in Britain in the winter, it may well not do so on a similar runway in the Middle East during the hot season where the air density and thus lift and thrust would be much reduced. In the latter case you may need to reach a higher speed on the runway at an increased flap setting and this may entail an increased take-off run due to the available engine thrust. The questions are, what aircraft weight? What speed? What runway length? What flap setting? What engine setting? What would be the stalling speed?

A further problem will occur when, for example, you attempt to assess the initial performance of a machine such as a gas turbine engine, or an aerofoil. If the original figures were read during a test in Britain at sea level during a cold weather snap we will have a problem if we attempt a similar test in the tropics at a thousand feet above sea level. The problem here is that the latter test figures may reveal a loss of performance but you may be unable to judge if the loss is due to the change in air density alone or a combination of low air density and mechanical deterioration!

I hope you will now be aware that there is a need for an international standard set of figures for altitude, air density, air temperature and pressure to which we can apply corrections based on the prevailing ambient conditions at our location. Take, for example, a test carried out in Britain. We do the test, assess and record the performance against a standard set of ambient figures. When we wish to carry out the same test at another geographic location where the altitude and ambient conditions are different we take a note of the figures identifying the prevailing ambient conditions and also record the observed performance at these figures. We then use correction tables to add to, or subtract from the observed performance figure which will then reflect what we would have got if the test were carried under the standard ambient figures. We can now make a direct comparison between our corrected test figure and the recorded initial test figure. If they do not agree, then something other than air density is causing the difference.

Once we have a set of standard atmospheric figures we can employ correction tables to ascertain such things as correct performance, airspeeds, thrust settings and flap settings in relation to the prevailing atmospheric conditions anywhere in the World. I hope that these illustrated examples will have made you understand why we needed to establish the *International Standard Atmosphere (ISA)*. If we are to assess the performance of, or compare the performance of, machines that rely on air density for their operation then a standard is essential. At least our example aircraft in the Middle East would now hopefully lift off successfully!

CHAPTER ONE
PHYSICS OF THE ATMOSPHERE

Altitude (h)		Ambient Temperature (To)			Ambient Pressure (Po)		Speed of Sound (ao)		
Feet	Metres	K	Deg. C	Deg. F	lb/sq in	millibar	ft/sec	knots	m/sec
-1,000	-304.8	290.13	+16.98	+62.6	15.24	1050.4	1120.3	663.3	341.5
0	0	288.15	15.00	59.0	14.69	1013.2	1116.6	661.1	340.3
+1,000	+304.8	286.17	13.02	55.4	14.17	977.1	1112.6	658.8	339.1
2,000	609.6	284.19	11.04	51.9	13.66	942.1	1108.7	656.5	337.9
3,000	914.4	282.21	9.06	48.3	13.17	908.1	1104.9	654.2	336.8
4,000	1219.2	280.23	7.08	44.7	12.69	875.1	1100.9	651.9	335.6
5,000	1524.0	278.24	5.09	41.2	12.23	843.0	1097.1	649.6	334.4
6,000	1828.8	276.26	3.11	37.6	11.78	811.9	1093.2	647.8	333.2
7,000	2133.6	274.28	1.13	34.0	11.34	781.8	1089.3	644.9	332.0
8,000	2438.2	272.30	-0.85	30.5	10.92	752.6	1085.3	642.6	330.8
9,000	2743.2	270.32	-2.83	26.9	10.51	724.3	1081.4	640.3	329.6
10,000	3048.0	268.34	-4.81	23.3	10.11	696.8	1077.4	637.9	328.4
11,000	3352.8	266.36	-6.79	19.8	9.72	670.2	1073.4	635.6	327.2
12,000	3657.6	264.38	-8.77	16.2	9.35	644.4	1069.4	633.2	325.9
13,000	3962.4	262.39	-10.76	12.6	8.98	619.4	1065.4	630.8	324.7
14,000	4267.2	260.41	-12.74	9.1	8.63	595.2	1061.4	628.4	323.5
15,000	4572.0	258.43	-14.72	5.5	8.29	571.7	1057.3	626.0	322.3
16,000	4876.8	256.45	-16.70	1.9	7.97	549.1	1053.3	623.6	321.1
17,000	5181.6	254.47	-18.68	-1.6	7.65	527.2	1049.2	621.2	319.8
18,000	5486.4	252.49	-20.66	-5.2	7.34	505.9	1045.1	618.8	318.5
19,000	5791.2	250.51	-22.64	-8.8	7.04	485.6	1040.9	616.4	317.3
20,000	6096.0	248.53	-24.62	-12.3	6.75	465.6	1036.9	613.9	316.1
21,000	6400.8	246.54	-26.61	-15.9	6.48	446.4	1032.7	611.5	314.8
22,000	6705.6	244.56	-28.59	-19.5	6.21	427.9	1028.6	609.0	313.5
23,000	7010.4	242.58	-30.57	-23.0	5.95	409.9	1024.4	606.5	312.2
24,000	7315.2	240.60	-32.55	-26.6	5.69	392.7	1020.2	604.1	310.9
25,000	7620.0	238.62	-34.53	-30.2	5.45	375.9	1015.9	601.6	309.7
26,000	7924.8	236.64	-36.51	-33.7	5.22	359.9	1011.8	599.1	308.4
27,000	8229.6	234.66	-38.49	-37.3	4.99	344.3	1007.5	596.6	307.1
28,000	8534.4	232.68	-40.47	-40.9	4.78	329.3	1003.2	594.0	305.8
29,000	8839.2	230.69	-42.46	-44.4	4.57	314.8	998.9	591.5	304.5
30,000	9144.0	228.71	-44.44	-48.0	4.36	300.9	994.7	588.9	303.2
31,000	9448.8	226.73	-46.42	-51.6	4.17	287.4	990.3	586.4	301.9
32,000	9753.6	224.75	-48.40	-55.1	3.98	274.5	986.0	583.8	300.5
33,000	10058.4	222.77	-50.38	-58.7	3.80	261.9	981.7	581.2	299.2
34,000	10363.2	220.79	-52.36	-62.3	3.63	249.9	977.3	578.7	297.9
35,000	10668.0	218.81	-54.34	-65.8	3.46	238.4	972.9	576.1	296.5
36,000	10972.8	216.83	-56.32	-69.4	3.29	227.3	968.5	573.4	295.2
36,089	11000.0	216.65	-56.50	-69.7	3.28	226.3	968.1	573.2	295.1
37,000	11277.6	Ambient temperature remains constant from thispoint up to 65,617 ft.			3.14	216.6	Speed of Sound remains constant from this point up to 65,617 ft.		
38,000	11582.4				2.99	206.6			
39,000	11887.2				2.85	196.8			
40,000	12192.0				2.72	187.5			
45,000	13716.0				2.14	147.5			
50,000	15240.0				1.68	115.9			
55,000	16764.0				1.32	91.2			
60,000	18288.0				1.04	71.7			
65,000	19812.0				0.82	56.4			

Table 1.1 - The International Standard Atmosphere (ISA)

The International Standard Atmosphere table establishes mean figures for ambient temperature, ambient pressure, air density and even the speed of sound in air at different altitudes. It also establishes fixed altitudes for the tropopause, stratopause etc. As you already know, the tropopause in reality occurs at a higher altitude in the equatorial regions than in the Polar Regions. The ISA table identifies the tropopause as being at the mean altitude as found in NW Europe that is identified as being 11km (36,090ft).

I was not able to fit the changes in air density with altitude in Table 1.1 due to space constrictions so I will give you a shortened version of the ISA table below.

Altitude (ft)	Pressure millibar (Abs)	psi	Temp (°C)	Density (kg/m³)
0	1013.25	14.69	+15	1.225
5,000	843	12.23	+5.09	1.056
10,000	696.8	10.11	-4.81	0.905
15,000	571.7	8.29	-14.72	0.771
18,000	**505.9**	**7.34**	**-20.66**	
20,000	465.6	6.75	-24.62	0.653
25,000	375.9	5.45	-34.53	0.549
30,000	300.9	4.36	-44.44	0.458
35,000	238.4	3.46	-54.34	0.386
36,090	**226.3**	**3.28**	**-56.5**	
40,000	187.5	2.72	-56.5	0.302
45,000	147.5	2.14	-56.5	0.237
50,000	115.9	1.68	-56.5	0.186

Table 1.2 - ISA Table showing Air Density Variations

Do remember that providing the air temperature remains constant, the air density will be directly proportional to air pressure. If the air pressure is halved, so is the air density and vice versa.

There are some figures that you will definitely need to memorise from the ISA table. The ISA figures established for the mean sea level altitude are an air temperature of 15°C, air pressure of 1013.2 millibar (mbs), which is 101.3kPa or 14.7lb/sq in or 29.92 in Hg. The air temperature lapse rate up to the tropopause is 6.5°C per km or 1.98°C per thousand feet. The altitude of the tropopause is 11km (36,090ft) where the air temperature is -56.5°C. The air temperature above the tropopause remains constant up to an altitude of 20km (65,600ft). These are commonly used units but just to cover ourselves against an examiner who may try you out in different units I will give you a compendium of units for ISA sea level pressure, temperature and density.

ISA Sea Level		
Pressure(Abs)	Temperature	Density
14.69psi	15°C	0.077lb/cu ft
1013.25mbs	288.15K	1.225 kg/m³
101.325kPa	59°F	
29.92 in Hg (Mercury)		
760mm Hg (Mercury)		
1.013bar		
1 Atmosphere		

Table 1.3 - ISA Sea Level Conditions

Q Codes

One of the more obvious problems to be overcome when we try to relate atmospheric conditions to flying an aeroplane is that of knowing at what precise altitude we are. The instrument used to establish this is the altimeter. An altimeter can measure height above just about any chosen reference point. It is of little comfort to us if we are all using any convenient reference by individual choice. The problem is overcome by using three established references. QFE, QNH and Flight Level (QNE). This involves setting the altimeter to read height above one of the chosen references in cooperation with the local air traffic advisory service. This is done by adjusting the altimeter to an advised barometric pressure setting that is displayed in a small window in the instrument. The reading is shown in millibars in the UK but some American operators like to use inches of mercury (Hg). There is a proposal to alter this to Hectopascals in the future. For our purposes we will use millibars.

QFE refers to the aircraft's height above the airfield. When this is used, the altimeter will display the height above the airfield and will thus read zero feet when the aircraft is parked on the airfield. The air traffic controller will pass the barometric pressure at the airfield elevation to the aircraft and the pilot will set the altimeter accordingly. Sometimes the term QFE Threshold is used. This is the barometric pressure at the airfield elevation converted to that existing at the approach end of the airfield's runway. The altimeter will show height above the threshold and will read zero feet on touch down.

QNH refers to the aircraft's height above sea level. When this is used, the altimeter will display the height above the sea. When the aircraft is parked at the airfield at this setting, the altimeter will display the elevation of the airfield above sea level. The air traffic control will advise the pilot to set the altimeter to a reading that is the airfield barometric pressure adjusted to that of local sea level.

Above a point known as the transition altitude, normally around 6,000ft in the UK, higher in the US, all aircraft set their altimeters to a setting of 1013.2 millibars. This is called *Flight Level (QNE)*. This is in effect the barometric pressure existing at mean sea level. It may not reflect the actual sea level pressure for the day, it is in fact an ISA figure. This ensures that all aircraft operating above the transition altitude are using the same setting and thus there will be no confusion in establishing separation altitudes. If you hear that an aircraft is at Flight Level 80 for example, you will know that it is 8,000ft above mean sea level ISA. Though the code is QNE it is more often than not just referred to as Flight Level. All pilots put this setting in as a routine above transition altitude and not necessarily under instruction from air traffic control.

Another term, used by the military only, was QGH. In case a question is set by an ex World War 2 pilot I will just make a mention of it for you. It is a barometric setting that will make the altimeter read height above a given altitude above the airfield from which a safe approach can be made. It was used to enable military aircraft to descend through low cloud cover and not hit the ground. Some civil airports may still retain this code for military aircraft diversions etc. You are unlikely to be asked about it.

Pressure Altitude

The International Standard Altitude table gives mean barometric pressures for given altitudes. We know that in reality the actual pressure at any altitude will vary from day to day or even hour to hour. It will also alter geographically. Say, for example, we were flying at an actual altitude of 5,000ft above mean sea level (amsl). The ISA barometric pressure for 5,000ft is 843 millibars. Let us assume that we have a prevailing low-pressure weather system at the time and that the recorded barometric pressure at 5,000ft was actually 811.9 millibars. This pressure corresponds to an altitude of 6,000ft on the ISA table. We need to be able to identify which is which. We could say that we were flying at an altitude of 5,000ft amsl but that we have a *pressure altitude* of 6,000ft.

The altimeter reads altitude in feet not barometric pressure. The instrument is calibrated to display the reading in feet in response to ISA barometric pressure variations. If the altimeter is set to Flight Level with 1013.2 millibars in its subscale window the altimeter will be displaying the pressure altitude 6,000ft not the actual altitude 5,000ft. The reassuring bit is that all other aircraft in the area will also be reading the pressure altitude, as they will be operating on the Flight Level setting.

CHAPTER ONE
PHYSICS OF THE ATMOSPHERE

Speed of Sound in Air

The speed that sound travels in air is 331 m/s at sea level when the air temperature is 0°C. Sound travels four times faster in water and fifteen times faster in steel. From this you would imagine that air density would affect the speed of sound in air. Strangely, it does not. The reason is found in the relationship that the velocity of sound in air has with the ratio of air pressure and density. The velocity of sound in air is proportional to the square root of air pressure divided by the air density. If the air pressure were to rise then so also would the density. In fact, if you doubled the air pressure, the density would also double. The ratio does not alter. However, this ratio only holds at constant air temperature. If the air temperature were to reduce at constant pressure it would increase the air density without altering its pressure. The ratio then changes.

If we look at the relationship again we would see that the speed of sound in air would reduce with a drop in temperature and increase with a rise in temperature. Put simply, *the Speed of Sound in Air is Proportional to the Absolute Air Temperature*. Air temperature falls by 1.98°C for every 1,000ft increase in altitude up to the tropopause, so it follows that the speed of sound in air will also reduce as altitude increases up to the tropopause. Remember, it is only the air temperature that affects the speed of sound in air.

Many commercial aircraft carry mach meters. The Mach No. is a unit that relates the true airspeed of an aircraft to the local speed of sound in air. The figure is derived by dividing the aircraft's true air speed by the local speed of sound in air. For an aircraft climbing at constant airspeed the gradual reduction in the value of the speed of sound in air means that the aircraft's mach meter will record a gradual increase in the mach number displayed. This increase is a true indication of the aircraft's increasing Mach speed.

Airspeed

Another problem encountered because of the change in air pressure and density is in establishing how fast an aircraft is flying. The instrument used to indicate this is the air speed indicator. This instrument uses a measure of the dynamic pressure in the air. Dynamic air pressure is directly affected by air density. If the air density falls, so does the dynamic pressure. As we know, air density reduces with increasing altitude, which means that at altitude the air speed indicator is not going to show the true speed of the aircraft, it will under read. The reading on the indicator is called the *Indicated Airspeed(IAS)*. It is quite easy to calculate the *True Airspeed(TAS)* providing we know the original air density prevailing when the indicator was calibrated and the density of the air at the height we are at. Naturally, you are not expected to carry out this calculation. Modern flight systems can do this for you. We will not concern ourselves with the calculation, as it is a Module 2 - Physics task. You need to remember that there are two airspeeds, IAS and TAS and that *IAS is less than TAS at altitude*. For example, if you were flying at 6,000ft at an IAS of 204kts (105m/s) your TAS would be typically 278kts (143 m/s).

14

If you have worked through the above in one sitting you may wish to take a coffee break before trying your hand at the revision questions that follow this. This will give you an idea of the style and depth of questions you will encounter in the various examination centres and of course your powers of recall.

CHAPTER ONE
PHYSICS OF THE ATMOSPHERE

Revision

Physics of the Atmosphere

Questions

1. The altitude of the Tropopause according to the International Standard Atmosphere is:

    a. 18,090ft

    b. 36,090ft

    c. 56,090ft

2. As the humidity of the air increases its density will:

    a. increase

    b. remain constant

    c. decrease

3. The oxygen content of the lower atmosphere is:

    a. 21% by volume

    b. 28% by weight

    c. 78% by volume

4. At what altitude does the Stratosphere commence according to the International Standard Atmosphere?

    a. 62,000ft

    b. 11km

    c. 22km

5. The barometric pressure at sea level ISA is:

    a. 1013.25mb

    b. 0.1kPa

    c. 101.325mb

6. The rate of decrease of air density with increasing altitude in the Troposphere in comparison to that in the lower Stratosphere is:

    a. faster

    b. slower

    c. similar

7. A millibar is a unit of:

    a. air density

    b. pressure altitude

    c. barometric pressure

8. The temperature in the lower Stratosphere:

    a. increases with altitude

    b. decreases with altitude

    c. remains constant

9. As altitude increases from sea level to the Tropopause the ratio of the percentage volumes of oxygen to nitrogen in the atmosphere:

    a. increases

    b. remains the same

    c. decreases

CHAPTER ONE
PHYSICS OF THE ATMOSPHERE

10. QNH refers to height above:

    a. an airfield

    b. sea level

    c. a safe approach altitude

11. The percentage of water vapour held in the air will:

    a. increase with an increase in air temperature

    b. decrease with an increase in air temperature

    c. be unaffected by air temperature

12. The ISA value for the temperature at the Tropopause is:

    a. 56.5°C

    b. -15°C

    c. -56.5°C

13. The barometric pressures quoted for the International Standard Atmosphere are:

    a. absolute

    b. gauge

    c. dynamic

14. The Q codes for barometric pressure corrections are:

    a. QEF, QEN, QFI

    b. QNH, QFE, QNE

    c. QHN, QHG, QIF

15. **The term Flight Level 50 means:**

    a. 50,000ft above mean sea level

    b. 5,000ft above airfield elevation

    c. 5,000ft above mean sea level

16. **The value of the speed of sound in air is:**

    a. proportional to the absolute air temperature

    b. inversely proportional to the air temperature

    c. proportional to air density

17. **If air temperature remains constant its density is:**

    a. inversely proportional to the barometric pressure

    b. directly proportional to the barometric pressure

    c. remains constant with variation in barometric pressure

18. **As altitude increases up to the Tropopause the air temperature:**

    a. does not decrease at a constant rate

    b. does decrease at a constant rate

    c. increases exponentially

19. **The indicated airspeed IAS at altitude will be:**

    a. higher than the true air speed

    b. the same as the true air speed

    c. lower than the true air speed

20. If an aircraft is flying at a given true altitude when the barometric pressure is below the ISA value for that altitude the indicated pressure altitude will be:

   a. higher

   b. lower

   c. the same

21. If a gauge pressure of 20psi is observed under standard ISA sea level conditions, the absolute pressure is:

   a. 5.3psi

   b. 34.7psi

   c. 14.7psi

22. The instrument used to measure absolute pressure would be a:

   a. manometer

   b. hydrometer

   c. barometer

23. At what altitude would the barometric pressure be half that of sea level pressure under ISA conditions:

   a. 18,000ft

   b. 33,000ft

   c. 12,000ft

24. If an aircraft is climbing at a constant true airspeed its Mach No. will:

   a. not alter

   b. decrease

   c. increase

25. The density of air at sea level ISA conditions is:

   a. 1.056kg m^3

   b. 1.225kg m^3

   c. 0.225kg m^3

Revision

Physics of the Atmosphere

Answers

1.	B	21.	B
2.	C	22.	C
3.	A	23.	A
4.	B	24.	C
5.	A	25.	B
6.	B		
7.	C		
8.	C		
9.	B		
10.	B		
11.	A		
12.	C		
13.	A		
14.	B		
15.	C		
16.	A		
17.	B		
18.	B		
19.	C		
20.	A		

Aerodynamics

Airflow Around a Body

Air is a compressible fluid that has low viscosity. For these reasons air is able to flow and change its shape easily in response to the smallest of pressure variations. Water is also a fluid and if you were to stand at the stern of a moving ship and look at the rough water in its wake you would not be surprised by the turbulence you saw there. You would not be so aware that the wake left by an aircraft passing through air is far more turbulent. The difference is, you cannot see the turbulence in air because it is invisible. The air must move to allow an object to pass through it. The airflow will initially try to follow the contours of the object as a streamlined flow but it will have a tendency to break away and swirl into fierce eddies or vortices. These will finally dissipate in the wake of the object. Just how turbulent the airflow becomes and how soon this will occur will depend on the shape of the body and the smoothness of its surfaces.

Before we examine how air reacts to the passage of some basic shaped objects we need to define exactly what is meant by the term *streamlined flow*. Imagine that we have an object mounted in a wind tunnel and we are making the air pass over it. You cannot see the airflow so we will make it visible by injecting multiple jets of smoke into the upstream flow. Where the airflow is undisturbed well in front of the body the smoke jets will appear to you as parallel lines. These are called streamlines. Imagine that the air now deflects to pass around the body. If the parallel streamlines remain separated as they pass over the body the airflow is described as being streamlined. The degree to which a body is said to possess a streamlined shape depends on how long the flow can remain in streamlines. Ideally, the streamlines will be maintained into the wake of the body. We are talking here of a perfect shape of course, in reality there will come a point in the flow across a body where the streamlines will start to intermingle as the flow breaks up into turbulence that will continue on to produce vortices in the wake.

At this point it would be useful to examine the flow across a number of basic shaped objects.

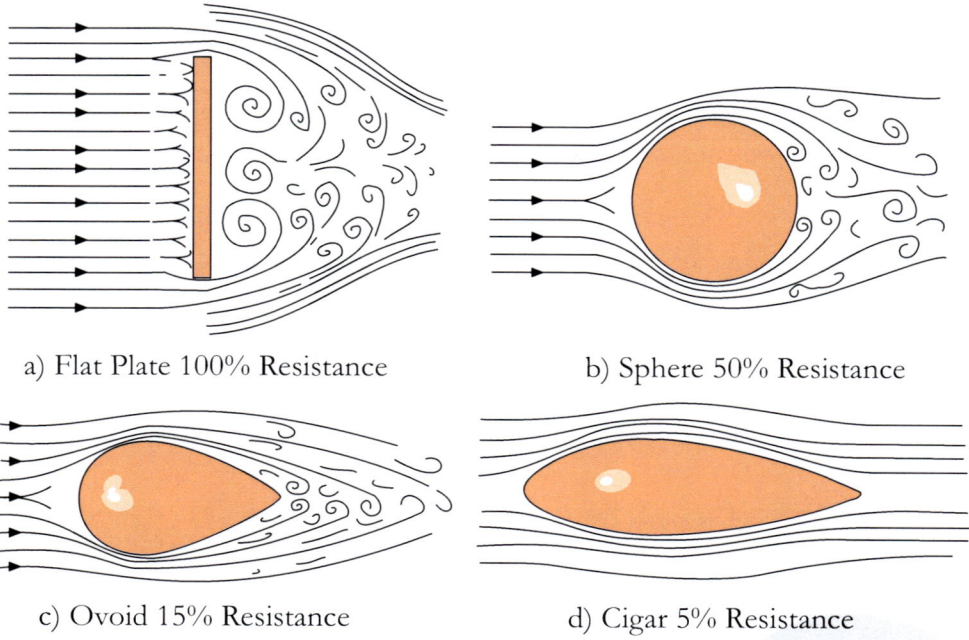

a) Flat Plate 100% Resistance b) Sphere 50% Resistance

c) Ovoid 15% Resistance d) Cigar 5% Resistance

Figure 2.1 - Affect of Shape on Streamlined Flow

Observe that in fig 2.1a most streamlines hit the front of the flat plate whilst others squeeze around the edges and immediately break up into extreme turbulence as the flow re-circulates into fierce vortices behind the plate. The air striking the front of the plate loses its kinetic energy, creating a pressure that will act on the plate to produce a force trying to push the plate rearwards. Imagine that if the plate were moving through still air, this force would produce a big resistance to the passage of the plate through the air. This is most definitely not a streamlined shape.

Now look at fig 2.1b and imagine it to be a cannon ball. The streamlined flow ahead of the ball senses the presence of the body and separates to pass by it. The flow attempts to stay attached to the ball's surface and maintains a streamlined flow until it tries to reunite in the wake. It cannot do it and breaks up into a turbulent re-circulating flow behind the ball. The kinetic energy given up by the airflow as it meets the ball is less than that lost on a flat plate with the same cross-sectional area. There will also be some frictional force produced by the airflow in contact with the ball's surface and a further drag force created as the air loses more kinetic energy by forming vortices and low-pressure regions in the wake. The total resistance force acting on a ball will be about half that of a flat plate having a similar cross sectional area. It is a good job cannons did not fire flat plates!

Now we are going to get a bit clever and use the egg shaped object in fig 2.1c. The correct description of this shape is ovoid. Now the streamlines deflect a lot more smoothly and remain separated over most of their journey over the body. But again they will break up as they experience difficulty in re-uniting in the wake. The total resistance force felt by this shape and the turbulence created will be significantly less than the previous two shapes but we are still not really streamlined.

Fig 2.1d is a different matter. It is a cigar shaped object. If it is perfectly dimensioned and ultra-smooth, the streamlines will remain intact throughout their passage over it. This is a streamlined shape. The best dimensions for such a shape are that the length of the body should fall somewhere between 3 and 4 times its maximum thickness which should occur at a third of the length back from the leading edge. This is called the ideal *Fineness Ratio* and if applied to this shape it gives the optimum streamlining and creates the least resistance to the body's passage through the air. Do not be surprised to learn that many species of fish had worked this out millions of years ago! Boat and airship designers also use this ratio as an optimum standard for low resistance. We will re-visit this later in the chapter.

Boundary layer

When air flows over a body there will be contact between the airflow and its surfaces. The air directly in contact with a surface will experience friction and will virtually be slowed to a stop. This contacting layer will also attempt to stop the flow of the layer of air just above it. Air has an extremely low viscosity so the surface layer can only retard the progress of the next layer out, not stop it. This means that the second layer will shear across the surface layer. Each successive layer of air moving out from the surface of the body will experience retardation to a lessening degree until at a given distance from the surface the airflow velocity will have been fully restored. This means that there will be a layer of airflow over the body a few centimetres thick that is made up of several thinner layers, each maintaining a slightly different velocity by shearing across each other. This layer is called the boundary layer. The thickness of the boundary layer and the frictional forces created by it depend on the smoothness of the surface of the body it contacts. There is no exact edge to a boundary layer, its influence merely fades away into the main airflow stream. The frictional forces created by the shearing in the boundary layer are collectively referred to as *Skin Friction*. A strong wind blowing along an airfield's runway can create a boundary layer on it up to two feet thick. You can get some idea of the velocity of the inner contacting layer on a light aircraft wing. If the wing had collected dust in the hangar you will find that much of it will still be evident on the wing after it has flown.

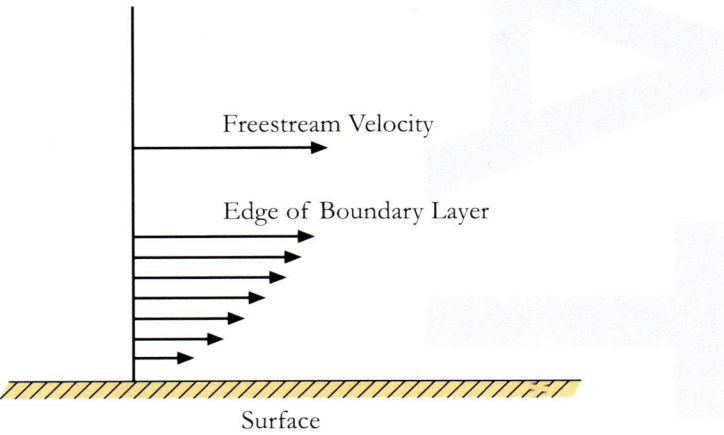

Figure 2.2 - Boundary Layer

Laminar & Turbulent flow

There are two types of boundary layer flow. The boundary layer will commence at the leading edge of a body. If the body is streamlined and smooth, the layers of air within the boundary layer will initially slide over each other in a streamlined manner. This is called *laminar* flow. As the air flows back across the body the shearing action between the layers produces tight vortices that develop until they cause the layers to break away and become turbulent. The point on the surface of the body where this occurs is called the *Transition Point*. It is the point where the laminar flow beaks up into *Turbulent Flow* and the boundary layer will thicken significantly. There is one point that you will need to remember. The transition point will occur earlier as airspeed increases. If an aircraft increases its airspeed, the transition point on its wings will move forward.

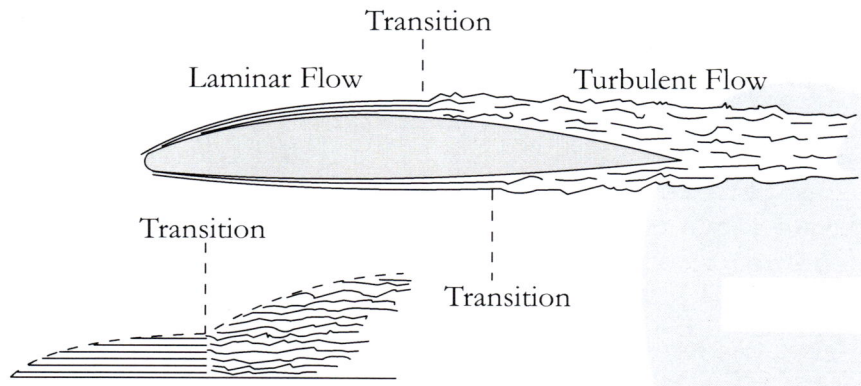

Figure 2.3 - Transition Point

In the outside chance some examiner thinks of it I will mention the Reynold's No. This number is determined by the surface length, density of the air, its velocity and its viscosity. Each type of surface will have a given number based on the above specifications. It is just a number and has no units just like a ratio. When this number is reached, the airflow becomes turbulent. It is a useful guide to understanding what causes turbulent flow. The number may be calculated using the following formula.

$$\text{Reynold's No.} = \frac{\rho v l}{\mu}$$

Where ρ is air density, v is air velocity, l is the distance from the leading to the trailing edge and μ is the coefficient of viscosity.

You can play with this formula. For example, if air density ρ, length l and the coefficient of viscosity μ remain constant, an increase in air velocity will increase the Reynold's No.

The Reynold's No. applies to all types of fluid flow. I think the operators of the trans Arabian oil pipeline would wish to adjust their oil flow to stay under the number specified for their installation.

To sum up, Laminar flow occurs when the thin layers of air in the boundary layer slide relatively smoothly over each other and maintain their separation. Turbulent flow occurs when the thin layers in the boundary layer merge and break up into a combination of rotational and linear movements. Laminar flow changes to turbulent flow at the transition point. Transition is delayed if the surface is smooth and even. The transition point will move forward with increasing airspeed. The boundary layer thickens with turbulent flow and skin friction increases.

Boundary Layer Separation

Later in this chapter you will read that a region of low pressure exists on top of a wing. This is created by the increased air velocity over the cambered top surface and is deepest at the point of maximum air velocity. This occurs at a point about one third of the chord length back from the leading edge. This means that the air passes back from this point into a region of recovering or adverse pressure that tries to slow it down. If the deceleration is too drastic, the boundary layer can separate from the wing. As the angle of attack of the wing increases, the adverse pressure gradient also increases and the lower part of the boundary layer on the wing will slow to a point where it tends reverse direction and move forwards, attracted by the low-pressure region ahead of it. A *Separation Point* will be reached where the boundary layer will separate completely from the wing and sweep around into a turbulent, re-circulating flow. Lift will be destroyed and drag will increase dramatically. You may already have recognised that this is what is known as a stalled condition and it occurs at an angle of attack of around 15°. I apologise for referring to some terms that have yet to be defined in this chapter but the close connection with the boundary layer makes it appropriate to introduce you to the topic of separation here.

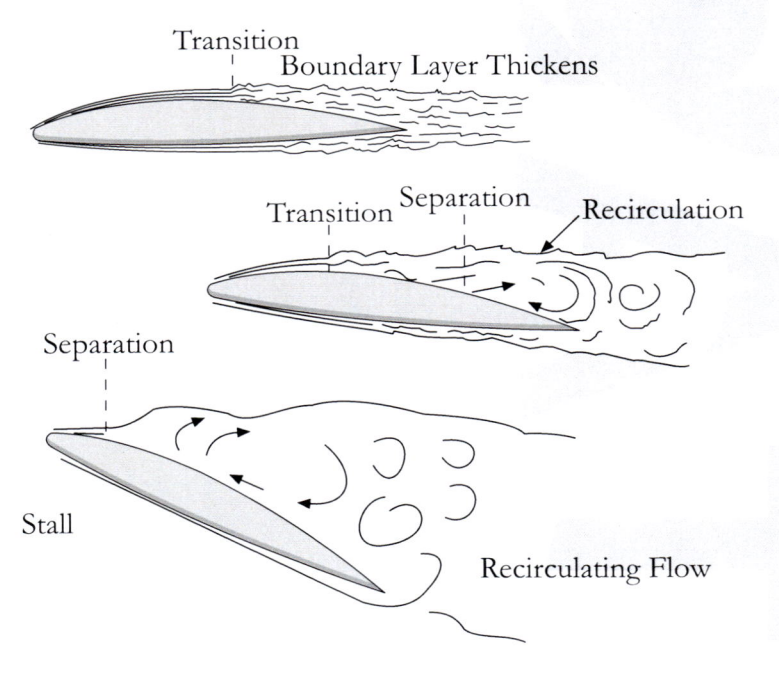

Figure 2.4 - Boundary Layer Separation

Free Stream Flow

Free stream flow may be defined as the air flowing in a region where its pressure, velocity and temperature are not affected in any way by the passage of an aircraft through it. The term **Relative Airflow (RAF)** is often used to describe free stream flow.

The Aerofoil

We have been discussing a number of body shapes and the behaviour of air flowing over them. I would now like to introduce you to a shape that is used in the construction of lifting bodies like an aircraft wing for example. Prior to this we need to examine a few of its underlying principles.

When air flows at subsonic velocities through the venturi shown in figure 2.5 there will be an interchange in its potential and kinetic energies. The sum of these two energies will however remain constant throughout. The potential energy of the air is simply its static pressure (p_{amb}). The kinetic energy of the air is the energy it possesses by virtue of its movement and is referred to as its dynamic pressure ($½\rho v^2$) where ρ is the air density and v is the air velocity. Dynamic pressure is often just called Q pressure.

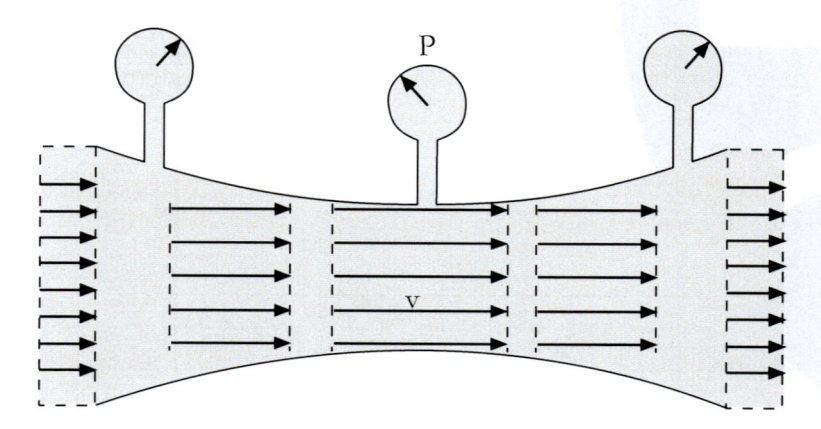

Figure 2.5 - Incompressible Flow

To explain the behaviour of air flowing through the venturi I think you would benefit by examining the Equation for the Continuity of Mass Airflow.

Density(ρ) × Cross-sectional Area(A) × Velocity(v) = A Constant(k)

Now look at the air flowing into the left hand side of the venturi. It is flowing through a converging duct so the cross-sectional area (A) is reducing. We assume the density (ρ) remains constant because we are not compressing the air, so the velocity (v) must increase to keep our equation balanced. If the air velocity increases then its dynamic pressure ($\tfrac{1}{2}\rho v^2$) must increase also. As the sum of static and dynamic pressures do not change then the static pressure (P_{amb}) has to reduce. Notice how the static pressure gauge in figure 2.5 indicates a fall in pressure. The air velocity will reach a maximum value in the throat of the venturi and the static pressure will then be at its lowest value. The air temperature will have fallen to its lowest value also.

Now look at the air flowing out of the right hand side of the venturi. The cross-sectional area (A) is increasing so the air velocity (v) must reduce. This means the dynamic pressure will reduce so the static pressure will increase again, as will the air temperature. In theory the air will have the same energy at exit as it had on entry. Notice how the static pressure gauge indicates that the pressure has restored and that the air velocity has recovered.

We have actually just worked through **Bernoulli's Theorem** that states that 'The total energy contained in an incompressible flow of air through a venturi will remain constant'.

$$\text{Potential Energy}(mgh) + \text{Kinetic Energy}(\tfrac{1}{2}\rho v^2) = \text{A Constant}$$

or

$$\text{Static Pressure} + \text{Dynamic Pressure} = \text{A Constant}$$

If you have followed this so far we can now take a look at the aerofoil.

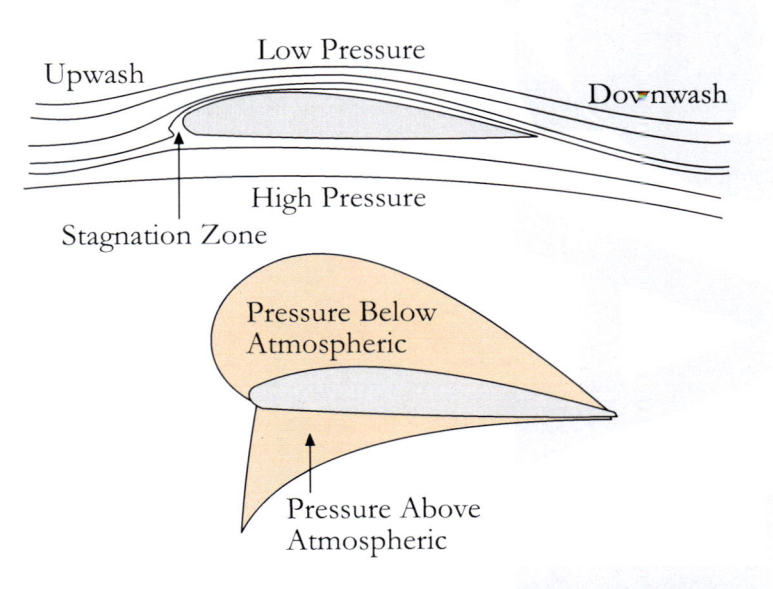

Figure 2.6 - Air Flow and Pressure Distribution around an Aerofoil

You have examined how air accelerates as it passes into the throat of a venturi. An aerofoil is shaped so that as it moves through the air its curved upper surface causes the air to accelerate over it and create a region of lowered static pressure. If the aerofoil is angled slightly to the direction of the free stream flow, the air striking the lower surface will move at a velocity that is slightly lower relative to the free stream velocity and this will create a region of higher pressure. If we now apply the aerofoil section to a wing, the low pressure region above the wing coupled with the higher pressure region below it creates a pressure difference that acts on the area of the wing and produces the upwards force we refer to as lift.

If you refer again to fig 2.6 you will see that the low-pressure region is more extensive than the higher-pressure region. The drop in pressure above a wing is greater and more extensive than the higher pressure beneath the wing. You could construe from this that an aircraft is sucked up into the air rather than pushed. In fact, around 70% of the lift force can be attributed to the low pressure created above the wing.

Looking again at fig 2.6, you will see that when the free stream airflow first encounters an aerofoil section wing it will divide just ahead of the leading edge. Notice that the air does not divide right on the nose of the leading edge but at a point forward of it and below it. The air is drawn upwards as it approaches the wing. This is known as *Upwash*. Look also at the streamlines. They close up as the air accelerates over the top surface and open out as the air decelerates underneath the wing. If the angle of the wing to the airflow is increased these velocity changes will also increase. Now take a look at the air flowing behind the wing. There is a downward flow of air that is referred to as *Downwash*.

Finally, if you look at the point where the airflow divides on approach to the wing there is a region where the particles of air must stop for a moment on making contact with it and prior to dividing. This region of stationary, or stagnating air is referred to as the *Stagnation* zone. The air pressure in this zone increases as the speed of the oncoming airflow increases. We will discuss these features in the following sections.

Upwash

If you will imagine for a moment, a solid, cylindrical body that is spinning around its axis as it moves through the air. Its spin direction is such that it moves the top surface away from the direction of travel and the bottom surface into the travel direction. Can you visualise that the air meeting the spinning cylinder will be drawn up and over the top surface and that the air in contact with it will accelerate. Now visualise the air trying to pass under the cylinder against the direction of spin. The contacting air will be slowed. It is as if you have a spinning vortex moving through the air if you imagine that the cylinder does not exist.

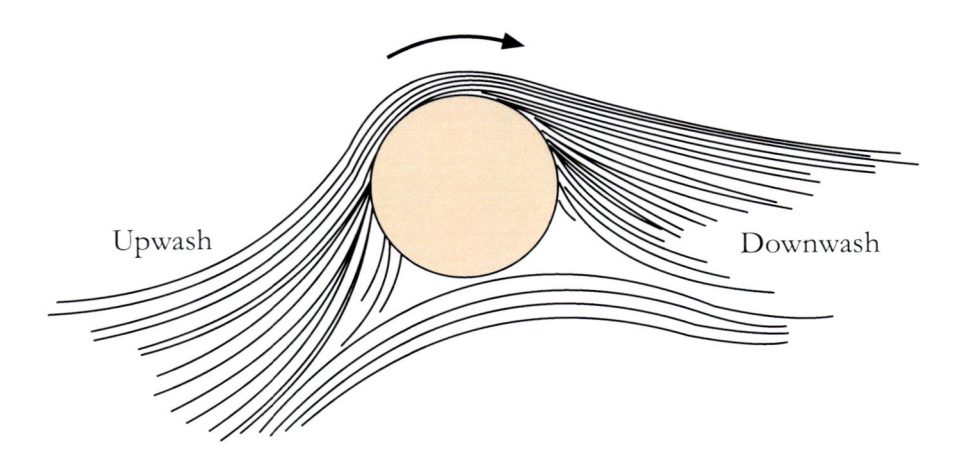

Figure 2.7 - Flow past a Spinning Cylinder

With your knowledge of the interchanges between kinetic and potential energies you will see that low pressure will occur at the top and higher pressure underneath. The spinning cylinder will actually create lift. This is known as the Magnus Effect. Golfers use it when they put spin on a golf ball.

The airflow over an aircraft wing resembles that of the spinning cylinder. The airflow over the top surface accelerates whilst that under the wing decelerates. Relatively speaking only, there appears to be a circulation around the wing, over the top and then forwards underneath. The effect is known as the wing bound vortex. That is purely a relative image of course. The result is that the airflow meeting the wing will be drawn upwards over the top just like it would if it met a spinning cylinder. That is a description of the **Upwash** experienced by the airflow encountering a moving wing.

Downwash

Imagine the spinning cylinder again. The spin will also throw the air passing over the top of the cylinder downwards as it tries to pass off the back. The wing will tend to do the same. This accounts in a small part only, for the **Downwash** of air off the back of a wing.

Aircraft wings have a low-pressure region over their upper surface and a higher-pressure region underneath. As the free stream air surrounding the aircraft is at atmospheric pressure and there are no material barriers between these regions of differing pressures something is bound to move. The surrounding air at atmospheric pressure attempts to move into the lower pressure region on top of the wings. This causes the airflow passing over the top surfaces to deflect slightly towards the wing root. The higher-pressure air underneath the wings tries to exit and join the lower atmospheric free stream air. This causes the air passing underneath the wing to deflect towards the wing tip. As the airflow leaves the rear of the wings, the upper and lower flows meet. Because they have been deflected in different directions they cross over each other at the trailing edge. This creates vortices in the airflow leaving the wing trailing edges.

At the wing tip, high-pressure air from under the wing spills out around the wing tip and curves up over the tip in an attempt to get into the low-pressure region over the wing. It can never do this because the aircraft is continuously moving. The result is that the air spilling out from under the wing tip spins forming a vortex. This vortex intensifies as it is joined by the trailing edge vortices that slide out to the tip to create an intense vortex at each wing tip, sweeping in behind the wing. Due to the movement of the aircraft through the air, the port and starboard vortices trail for several miles behind the aircraft descending all the time hence they are often called trailing vortices. Other descriptions used are, wing tip vortices or wake vortices.

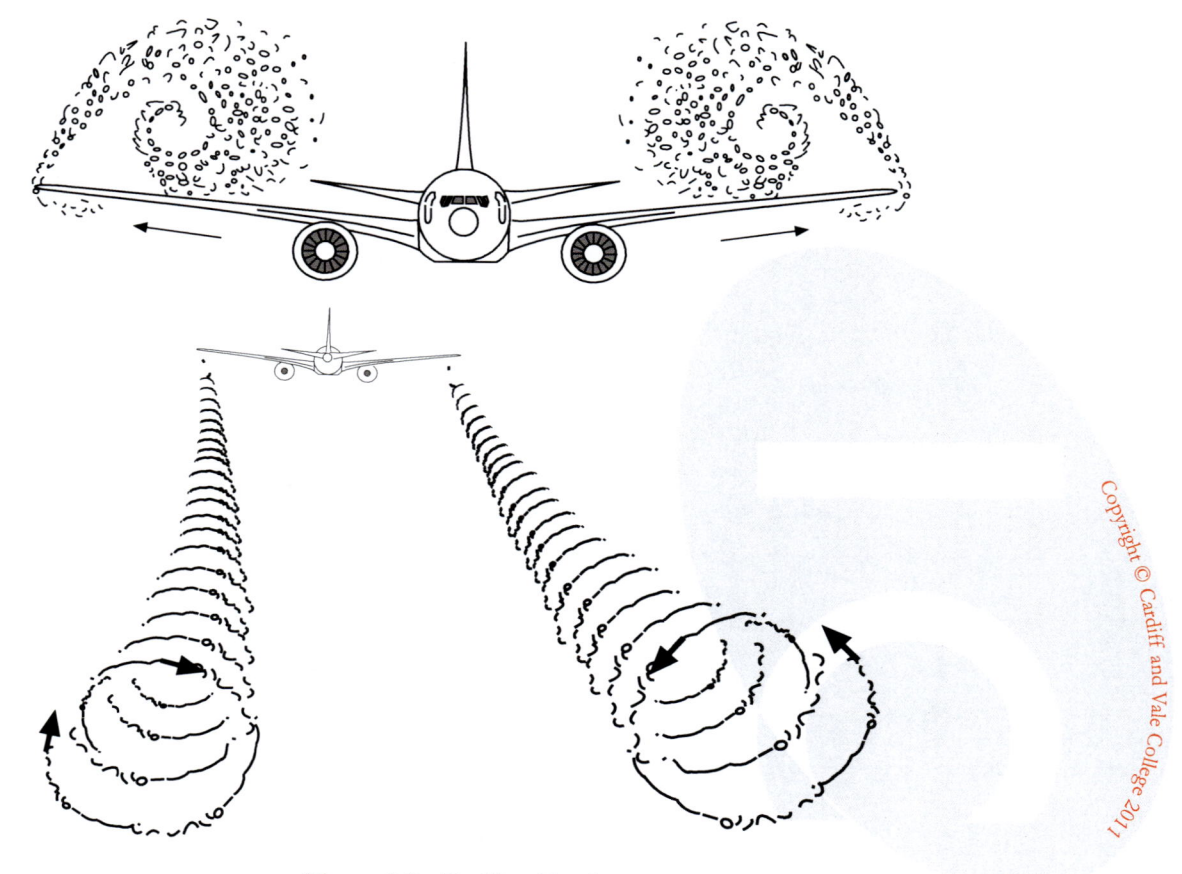

Figure 2.8 - Trailing Vortices

What has this all got to do with downwash? The spinning air off the wing tips curves over and exerts a downward push on the air leaving the wing trailing edges. The pressure difference between the upper and lower wing surfaces is related to the amount of lift they produce. When the lift produced by the wings increases the wing tip vortices intensify and the downwash of air leaving the wings increases. Aircraft wings produce most lift at low forward airspeeds when the wing angle of attack is high and when the flaps may be extended. This means that the downwash will be most evident during take-off, approach and landing and becomes less evident as airspeed increases. The aircraft weight, configuration of the wings and the chord width of the wing tips also have a bearing on the amount of air spilling over the tips. Light aircraft with narrow wings and small chord tips produce weak vortices and thus little downwash. For example, lightly laden gliders produce very little downwash. A large heavily laden commercial transport would produce a lot of downwash. Rearward swept wing aircraft experience a further vortex called a Ram's Horn Vortex that eventually joins the tip vortex and intensifies it. I will explain that later under the effects of swept wings.

Vortices

The definition of a vortex is a whirling motion of a fluid forming a cavity in the centre. The spin velocity of the fluid decreases with increasing radii from the centre. As your knowledge of energy exchanges now allows you can understand that the high velocity at the centre of a vortex will result in a very low static pressure there. The temperature at the core is also very low so any moisture in the air will precipitate out. This is why in humid conditions you can often observe the water vapour trails that mark out the path of a vortex core. Do not confuse these with the water vapour trails left behind the engines. They are produced by water vapour formed by the combustion of hydrocarbon fuels. The vapour trails I refer to are most readily seen streaming off the wing tips or the outer edges of the flaps in humid conditions.

We have already described the formation of the wing tip **Trailing Vortex** in the previous section on downwash. The effects of the downwash on the wing airflow are very significant. It alters the direction of the relative airflow to the wing and thus reduces the wing angle of attack. This has two effects. Firstly the lift will have reduced so the wing angle of attack has to be restored by tilting the wings to a higher angle to restore the lift. This will in turn increase the drag force. Secondly, because the lift force always acts at right angles to the relative airflow, the resultant lift vector is tilted rearwards and this induces another rearwards component of drag force. This latter effect is known as **Induced Drag** and is directly associated with lift production through the action of the trailing vortices on downwash. Simply, induced drag is highest at low airspeeds and reduces as airspeed increases. We will discuss this further later in this chapter.

Another source of vortex flow is connected to the **Wing Bound Vortex** that was described earlier in our discussion on upwash. This is the **Starting Vortex**. When a wing starts to move through the air the wing bound vortex is slow to develop but as it strengthens and lift starts to be produced a starting vortex forms at the trailing edge. It is equal to but opposite in sense to the circulation around the wing. During the acceleration experienced during take-off and when the aircraft is rotated on lift off, the vortex intensifies and then leaves the trailing edge, moving downstream behind the aircraft. You can demonstrate this if you dip a flat board into still water and then suddenly accelerate it forwards. You will see an eddy break away from the rear of the board. This vortex is parallel to the span of the aircraft wings and forms a closed loop together with the wing tip trailing vortices.

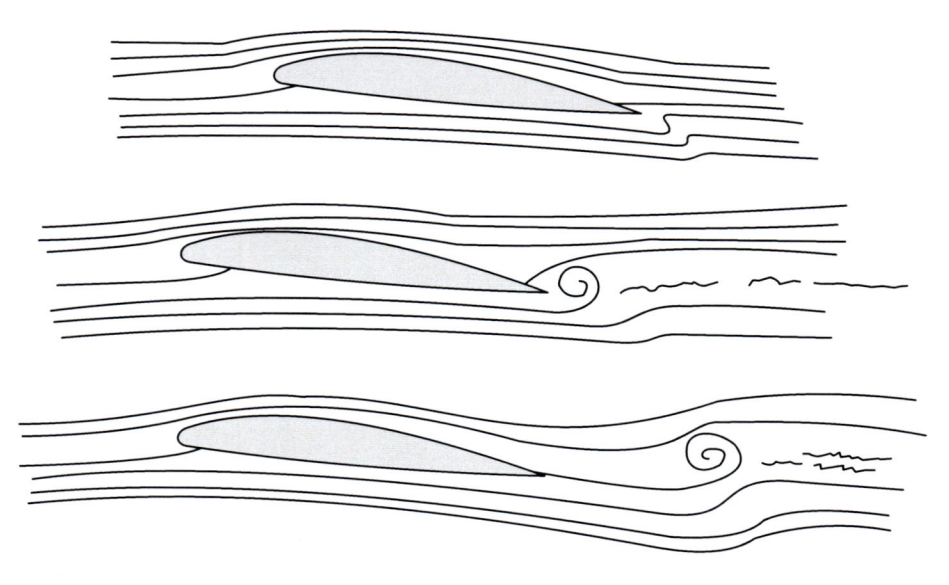

Figure 2.9 - Starting Vortex

The loop resembles a horseshoe shape and is referred to as the *Horseshoe Vortex* system. The starting vortex is left behind on the runway once the aircraft has lifted off. It can reappear when the aircraft pulls out of a dive for example. Again it will be left behind. There is a reversal of this vortex that forms and detaches during landing and this is referred to as the *Stopping Vortex*. Starting and stopping vortices may be formed during any sudden manoeuvres. They will dissipate in the air but can still be felt several miles behind the aircraft.

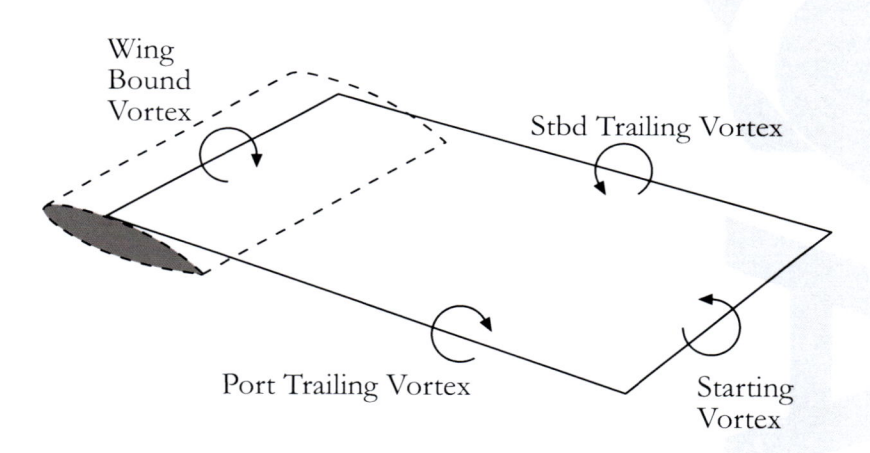

Figure 2.10 - Horseshoe Vortex

I have referred to the Ram's Horn Vortex in regard of swept wing aircraft. I will explain this under a separate heading dealing with the effects of sweepback.

There is one other vortex that is worth a brief mention in case a question finds its way to you. This vortex is associated with slender delta winged aircraft. The Concorde was one such example. Most conventional aircraft designs go to great lengths to avoid re-circulatory flow on the top of their wings. The slender delta wing on the other hand has a wing leading edge that is designed to cause the airflow to break away and spin into a deep re-circulating air vortex over the top of its wings. The reason for doing this is to take advantage of the low-pressure core of the vortex to produce additional lift on the slender delta. These vortices will exist during all the stages of the slender delta wing aircraft flight and is referred to as *Vortex Lift*. If you recall the Concorde had a significant nose up attitude at low speed that was much higher than that seen on conventional aircraft. This deepened the wing vortices to the extent that the lift coefficient and the lift/drag ratio of the slender delta were very high at low airspeed because of the additional vortex lift.

Stagnation

Air is said to stagnate when it is slowed to a virtual standstill. I have described this earlier in relation to the flow over an aerofoil. At the point the flow divides just under the wing leading edge, the airflow striking the surface stops prior to dividing. This creates a region of stationary air like a dead pool just in front of the underside of the wing leading edge. It is a high-pressure region and extends along the whole span of the wing leading edge. At high airspeeds the air in the stagnation zone becomes highly compressed and its temperature rises. For your interest only in this module it is this fact that prevents a blunt leading edge wing ever attaining supersonic flight. The shock wave that forms in front of the leading edge must attach itself to the edge in order that the aircraft can go supersonic. The fact that the air temperature in the stagnation zone increases means that the local speed of sound in air in the zone also increases and the wave can never attach itself to the leading edge. A sharp leading edge will have the effect of reducing the size of the stagnation zone to the point it is insignificant and even non-existent. You will encounter this topic again in the high-speed flight section of module 11.

Configuration & Aerodynamic Performance

Definition of Terms

In this section we are going to examine some of the more commonly used terms used in the description of the configuration and aerodynamic performance of aircraft and aircraft wings. We will do more than simply define them to save repetition later on.

Camber

The generation of lift by a conventional aerofoil section wing depends on producing low pressure over the upper surface and high pressure underneath. The high pressure may be created by merely presenting the wing at an angle to the relative air flow. Indeed you could conceivably make a paving slab produce lift by doing this. To generate the low pressure region, however, requires that the air be accelerated over the top surface of the wing. This is achieved by designing a convex, curved upper surface. The degree of curvature will influence the final velocity achieved by the airflow. In order that we may describe an aerofoil in such terms the word camber is used.

Maximum Camber

Mean Camber Line

Chord Line

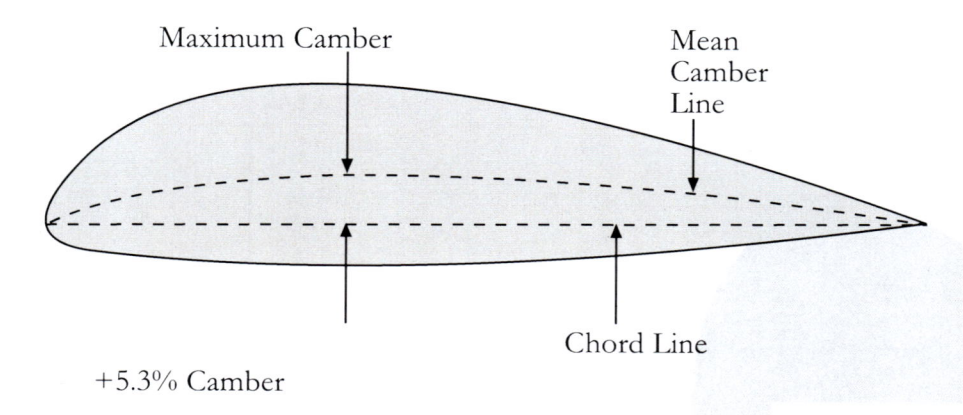

+5.3% Camber

Figure 2.11 - Camber

If you examine fig 2.11 you will see that a line is drawn that is equidistant from the top and bottom surfaces. This is the **Mean Camber Line**. Another line is shown that is straight and runs from the centre of curvature of the leading edge to the apex of the trailing edge. This is the aerofoil **Chord Line**. The camber of the aerofoil illustrated is measured as the maximum distance that exists between the mean camber line and the chord line. If the mean camber line is positioned above the chord line the aerofoil is said to possess **Positive Camber**. If the camber line is below the chord line it is said to possess **Negative Camber**.

Camber is often expressed as a percentage of the chord length. This necessitates dividing the maximum camber distance by the **chord length** and multiplying the result by 100.

$$\frac{\text{Maximum Camber}}{\text{Chord Length}} \times 100\% = \% \text{ Camber}$$

A high camber would be suited to low speed, high lift producing wings that have thick aerofoil sections. A low camber would be suited to high-speed aircraft wings that have thin aerofoil sections. Notice that in the illustration at fig 2.12, the symmetrical aerofoil has equally curved top and bottom surfaces but it has zero camber. This aerofoil can still produce lift but only when it is presented at a positive angle of attack to the relative airflow. It may be interesting to note that when an aircraft is flying inverted its wings can only produce lift by being presented at an angle of attack to the relative airflow.

Maximum Camber *Mean Camber Line*

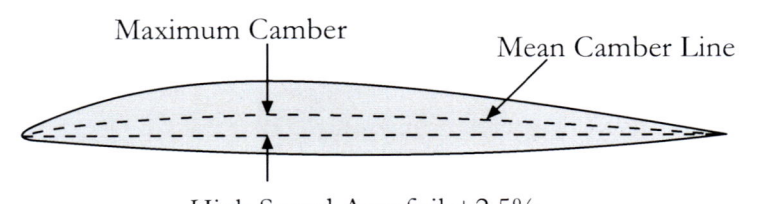

High Speed Aerofoil +2.5%

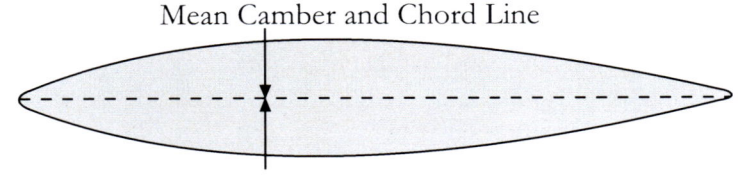

Mean Camber and Chord Line

Symmetrical Aerofoil 0%

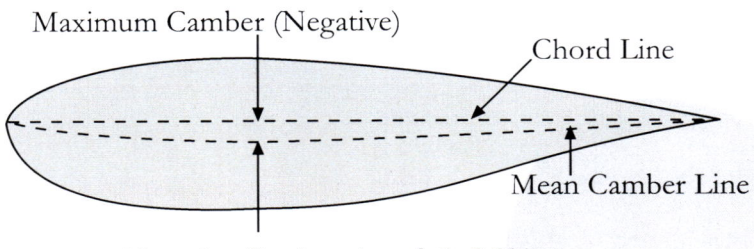

Maximum Camber (Negative)   Chord Line

Mean Camber Line

Negative Camber Aerofoil -3.3%

Figure 2.12 - Example Cambers

Chord

The chord of an aerofoil is an imaginary straight line that is drawn from the centre of curvature of the leading edge to the centre of curvature or apex of the trailing edge. Although rare these days, wings that may have a pronounced concave under-surface may have a chord line that travels outside the lower perimeter of the aerofoil. Most wings these days have either flat or slightly convex under surfaces.

Chord

Chord

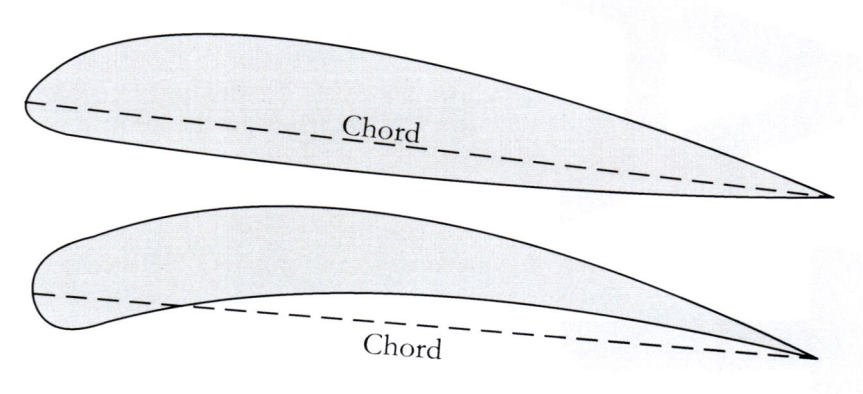

Figure 2.13 - Chord

Thickness/Chord Ratio

An aerofoil section may be described by its thickness/chord or t/c ratio. This is simply the maximum thickness of the section divided by the chord length. The result is usually multiplied by one hundred to give a percentage expression. A high t/c ratio for example 18% would indicate a thick-sectioned high camber aerofoil that would most probably give high lift at low airspeeds. A low t/c ratio for example 8% would indicate a slim longer chord aerofoil that would be more suited to higher airspeeds. The Concorde wings with their long chord lines and slender configuration had a t/c ratio of only 3%.

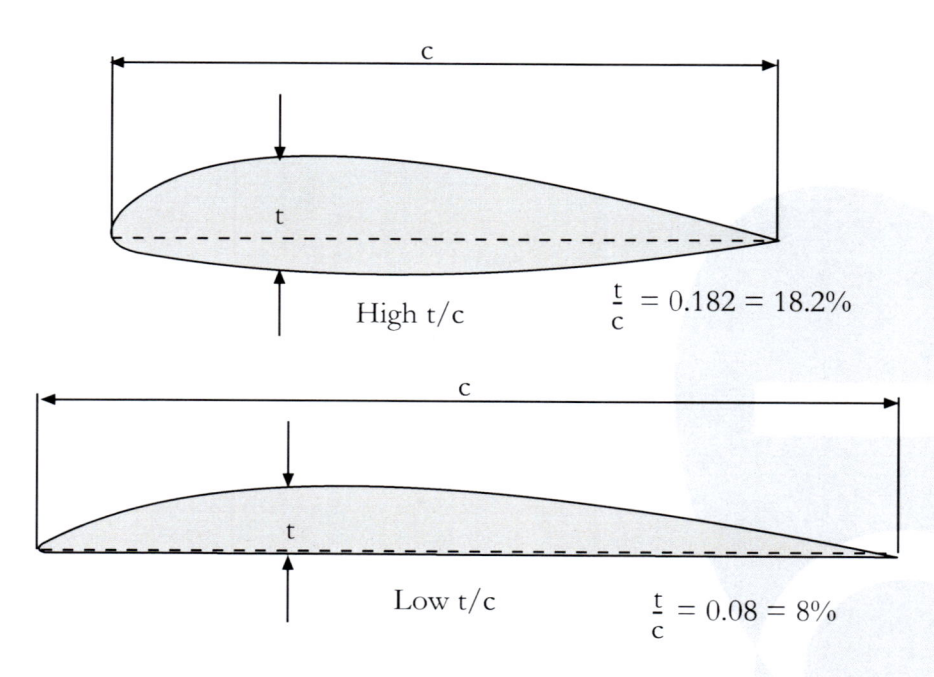

High t/c $\quad \dfrac{t}{c} = 0.182 = 18.2\%$

Low t/c $\quad \dfrac{t}{c} = 0.08 = 8\%$

Figure 2.14 - Thickness/Chord Ratio

Fineness Ratio

I have mentioned this earlier in this chapter. If you recall it is the aerofoil thickness to chord ratio. The ideal fineness ratio refers to a chord length that is between 3 and 4 times the maximum thickness of the aerofoil section. This represents a t/c ratio between 25% and 33%. From the illustration you can see that it would be impractical to maintain this ratio in all cases. Structurally it would be difficult to use it from wing root to tip in all but the simplest of low speed aircraft designs. It would not be suited to high-speed flight because it would be too thick, have too high a camber and too blunt a leading edge. It is an ideal not a rule.

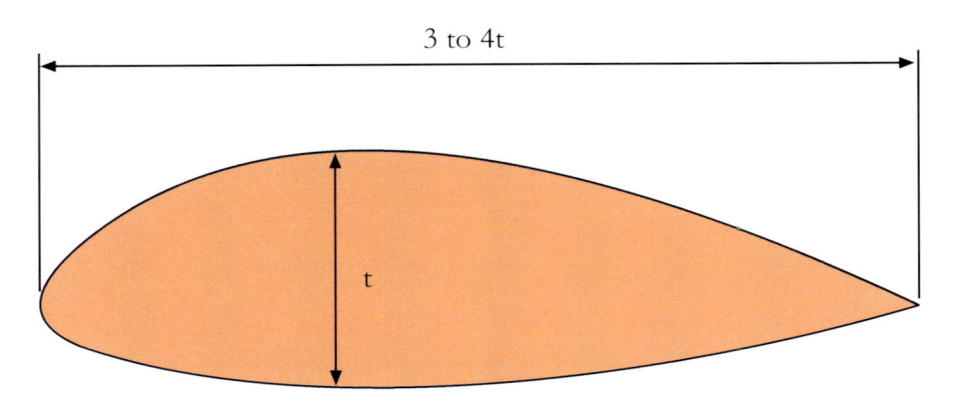

Figure 2.15 - Fineness Ratio

Mean Aerodynamic Chord

The mean aerodynamic chord is the average distance between the leading and trailing edges of a wing. Because wings often come in a variety of complex plan form shapes the chord length of the wing will vary from root to tip. The mean or average length is taken and the chord line on the wing that matches that length is identified as the mean aerodynamic chord line. Its position will be identified for each wing in all structural and maintenance manuals as a distance port or starboard of the aircraft centreline. The mean chord length can be calculated by dividing the wing area by the wingspan.

$$\text{Mean Chord} = \frac{\text{Wing Area}}{\text{Wing Span}}$$

When ascertaining the wing area we use the sum plan form area of both wings including the area where they cross the fuselage. The wingspan of an aircraft is the distance from wingtip to wingtip taken as a straight line between the two wing tips, crossing the aircraft at right angles to the fuselage centreline.

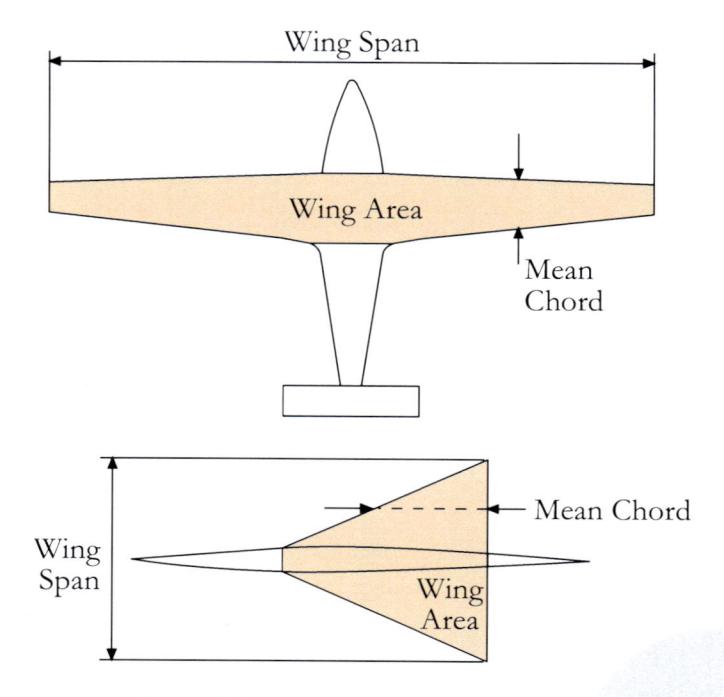

Figure 2.16 - Mean Aerodynamic Chord

Angle of Attack

The angle of attack (α) of an aerofoil section is the angle between its chord line and the relative airflow.

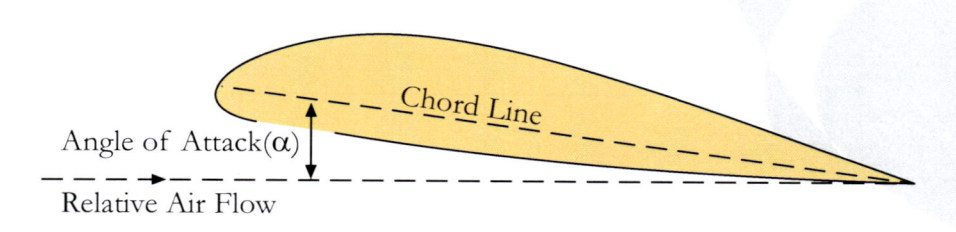

Figure 2.17 - Angle of Attack

As the angle of attack of a wing is increased its lift increases but so also do the drag forces acting on it. The most effective angle of attack for an aerofoil section wing is viewed to be where the lift force exceeds the drag forces by the largest multiple. This angle is between *Three and Four Degrees (3° and 4°)* angle of attack (AOA). At this position the lift force will be around twenty four times greater than the drag forces. As the angle of attack is increased from this value the lift/drag ratio reduces until at *Fifteen Degrees (15°)* the wing will stall and lift will be destroyed.

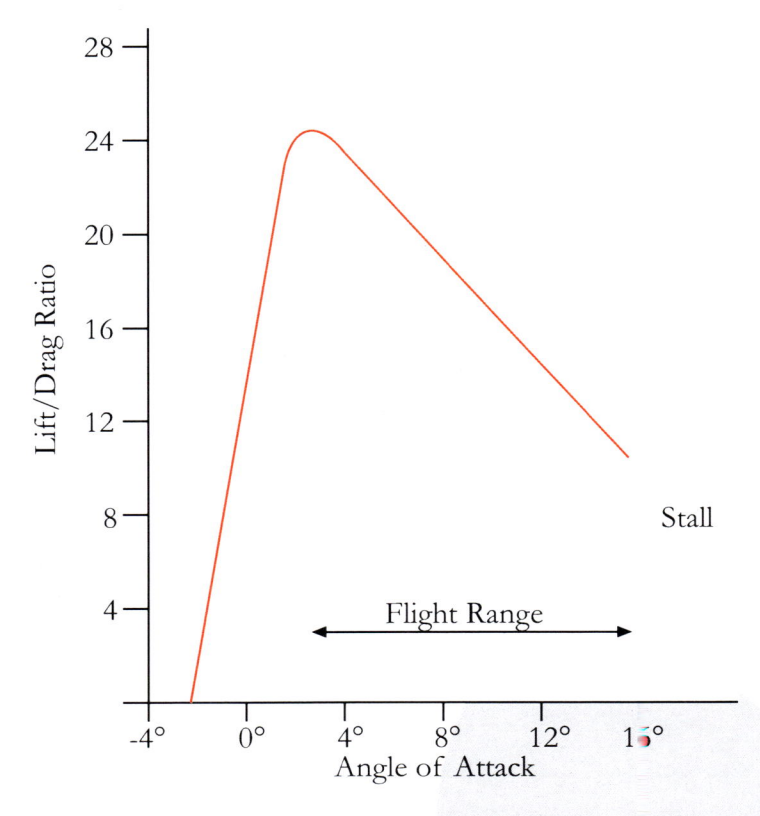

Figure 2.18 - Lift/Drag Curve Related to Changes of AOA

Angle of Incidence

The term angle of incidence is sometimes used to describe the angle of attack but this can cause confusion with what is known as the *'Riggers Angle of Incidence'*. The riggers angle of incidence is the angle at which the chord of an aircraft wing is set to the fuselage longitudinal datum, typically between 2° and 4°. It is a structural feature and has no connection with the angle of attack when used in this sense. You should be careful when faced with questions using both these terms and as a general rule use angle of attack to describe the angle between the wing chord and the relative airflow, and angle of incidence to describe the angle at which the wing is aligned to the longitudinal datum.

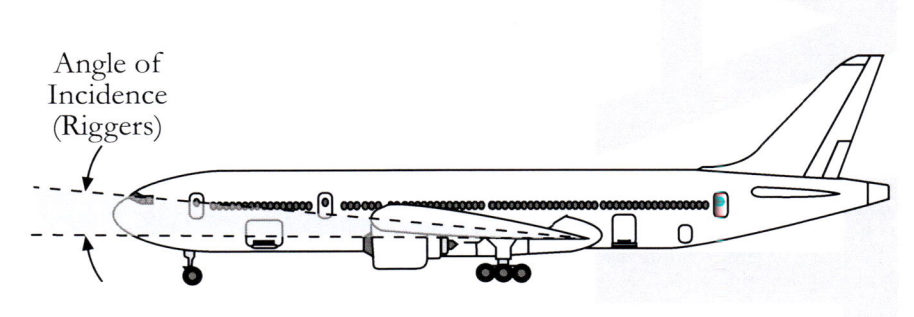

Figure 2.19 - Riggers Angle of Incidence

Wash in

This is a term used to describe the structural design of a wing that has a *Riggers Angle of Incidence that increases from the wing root to the wing tip*. In flight this would result in the wing experiencing an increased angle of attack and downwash towards the wing tips. If the aircraft were to approach the stall angle of attack, the wing tips would tend to stall first. As a general rule that is not a very desirable feature because it would create a stall right in the region where the aileron control surfaces are usually sited so the aircraft could lose lateral control at the onset of a stall. Further to this an increased angle of attack near the wing tips would tend to strengthen the wing tip vortices, increasing the down wash and as a result increasing the induced drag. Finally it could remove a useful stall warning namely the turbulence streaming off the outer wing areas at the onset of a stall would initially miss the tail-plane therefore the pilot would sense little or no buffeting just prior to the stall.

Wash in could, for example, be considered useful on a straight rectangular shaped wing that has a strong downwash near the tip that reduces towards the wing root. As the downwash reduces the angle of attack towards the tip this would result in a varying angle of attack and lift coefficient along the wing, reducing towards the tip. Wash in could be used to correct this but would in turn further intensify downwash and encourage tip stall. Not a good solution.

Wash Out

This term is used to describe the structural design of a wing that has a *Riggers Angle of Incidence that decreases from the wing root to the wing tip*. In flight this would result in the wing experiencing a decreasing angle of attack towards the wing tips. If the aircraft were to approach the stall angle of attack the wing roots would tend to stall first. This has the advantages already discussed. Lateral control can be maintained during the onset of stall. Stall warning would be available because the turbulence streaming off the wing root areas would buffet the tail plane as the stall is approached. Finally, the reduced angle of attack near the wing tips would reduce the tip vortices, reduce the downwash and reduce the induced drag. As a general rule it is more desirable that an aircraft stalls at its wing roots first for the reasons stated.

Centre of Pressure

The centre of pressure of an aerofoil is defined as a point on the chord line where the resultant of the lift forces produced by the aerofoil is said to act through.

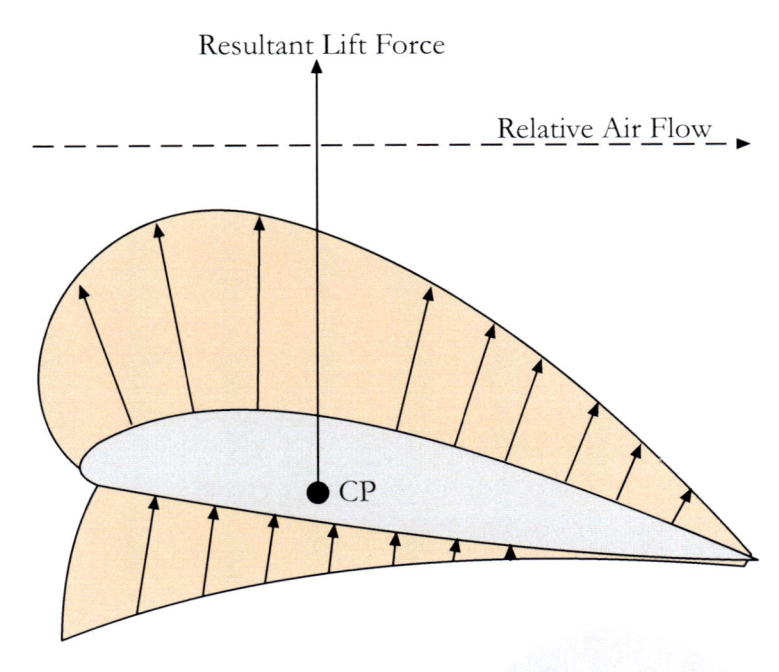

Figure 2.20 - Centre of Pressure

If the aerofoil is set at a given negative angle of attack, it will produce zero lift. As there is no lift the centre of pressure will not exist, it will theoretically be off the rear of the aerofoil at infinity. As the aerofoil angle of attack is increased through zero degrees to a positive angle the centre of pressure will move forwards along the chord line and will continue to do this as the angle of attack increases. As the angle of attack nears fifteen degrees the centre of pressure will have almost reached its fully forward position that is around 30% of the chord line from the leading edge. At fifteen degrees the fully forward position is reached. As the wing stalls the centre of pressure moves rapidly rearwards.

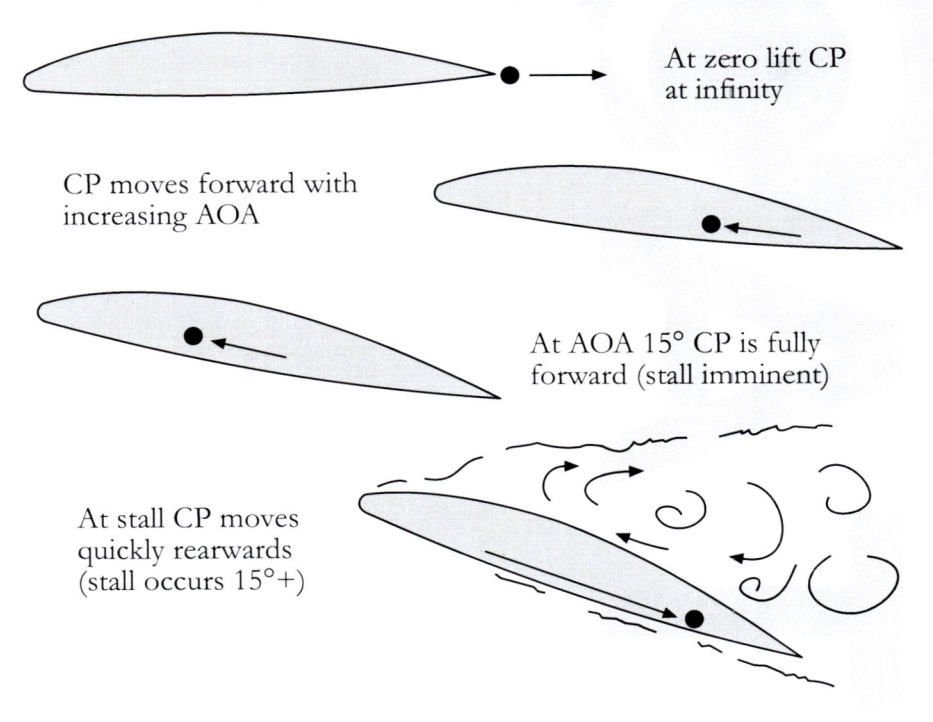

At zero lift CP at infinity

CP moves forward with increasing AOA

At AOA 15° CP is fully forward (stall imminent)

At stall CP moves quickly rearwards (stall occurs 15°+)

Figure 2.21 - Movement of Centre of Pressure

The movement of the centre of pressure causes changes to the pitch attitude of the aircraft because of its relationship with the position of the aircraft centre of gravity. During a stall condition the rapid rearwards movement of the centre of pressure creates a bigger moment arm around the C of G and the aircraft nose will drop quickly in the stall.

Aerodynamic Centre

There exists a point along the chord line about which there will be no alteration in pitching moment as a result of changes to the wing angle of attack. This is known as the aerodynamic centre of the wing and it is normally to be found at the quarter chord length from the leading edge.

Wing Shape and Aspect Ratio

This is the ratio of the aircraft wingspan to its mean chord length. This ratio is linked to the wing plan form configuration as opposed to its cross sectional configuration. The wingspan in this case is taken as the straight-line distance between the two wing tips.

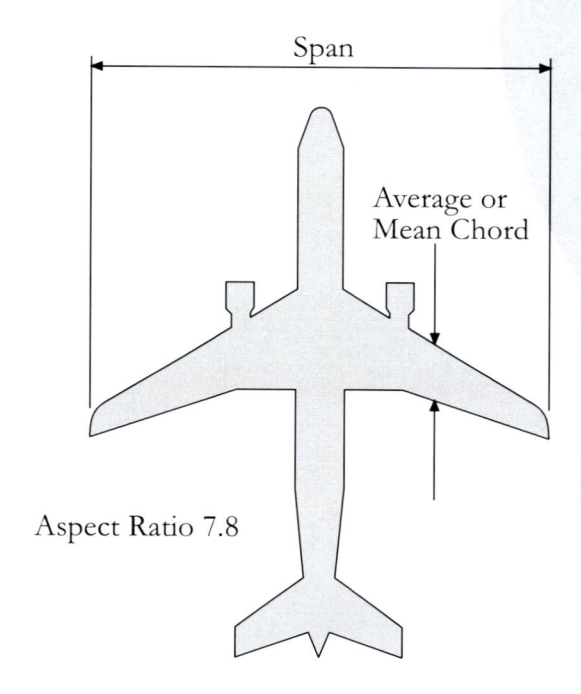

Figure 2.22 - Aspect Ratio

$$\text{Aspect Ratio} = \frac{\text{Span}}{\text{Mean Chord}} \text{ or } \frac{\text{Span}^2}{\text{Area}}$$

If you were to consider a rectangular wing having an area of say 24m² you can see that this same area can be arrived at using many different dimensions for span and chord. The wing could have a mean chord of one metre and a span of 24 metres. Again it could have a mean chord of 3 metres and a span of 8 metres. The former dimensions indicate a long narrow rather flexible wing while the latter indicates a shorter, broader and stiffer wing.

High aspect ratio wings with their wide span and narrow chord have a definite advantage in that they create low trailing vortices and thus minimise induced drag. Induced drag is inversely proportional to aspect ratio so they will have high lift/drag ratios. A very high aspect ratio would be perfect for a glider for example where low drag and high lift are very important. Unfortunately, the structural implications in terms of the wing bending moments of building a heavy commercial aircraft in that configuration make such dimensions impracticable. A balance has to be struck between structural strength and aircraft weight. It is here that the aspect ratio becomes an important feature of aircraft design and application.

The aspect ratio can be calculated in two ways. Either way will give the same ratio. In the first example you divide the wingspan by the mean or average chord length. In the second example you divide the square of the wingspan by the total area of the wings.

To take an example, consider a wing with a span of 10m and a mean chord of 2m.

$$1.\ \text{Aspect Ratio} = \frac{\text{Span}}{\text{Mean Chord}} = \frac{10\text{m}}{2\text{m}} = 5$$

$$2.\ \text{Aspect Ratio} = \frac{\text{Span}^2}{\text{Area}} = \frac{100\text{m}}{20\text{m}^2} = 5$$

High aspect ratios in the order of 20 and over would suit gliders or U2 spy planes for example. The middle range aspect ratios would suit many powered sub-sonic aircraft. The lower aspect ratios would be more applicable to higher speed, swept wing and delta winged aircraft. Military aircraft require agile manoeuvrability at high speeds and that demands structural strength and low aspect ratio wings.

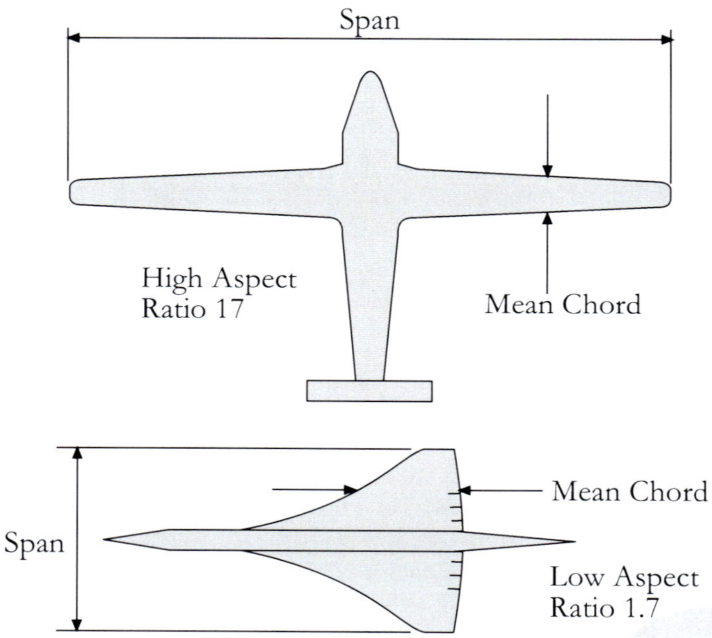

Figure 2.23 - High and Low Aspect Ratios

Wing Shape

Although we have discussed the general wing plan forms in relation to different aspect ratios this would not be complete without examining some of the configurations of wing actually found in use and their characteristics.

Taper Ratio

The taper ratio of a wing is the ratio between the wing tip chord and the chord at the wing root. A straight rectangular wing would have a taper ratio of 1. A wing with a tip chord half the length of the root chord would have a ratio of 0.5 whereas a wing that had a sharp pointed tip would have a taper ratio of 0.

Straight Rectangular

This configuration is normally for use on low speed aircraft. A feature of the straight rectangular wing is that it has a strong vortex at the tip with a correspondingly high downwash and a reduced angle of attack at the tips. The downwash is locally high at the wing tip reducing towards the wing root. Induced drag is high and the lift coefficient will vary along the wing due to a variation in angle of attack from wing root to tip, low at the tip and higher at the root. This results in uneven loading on the wing, the tip producing less lift than the root. The straight rectangular wing is less efficient than an elliptical wing. A high aspect ratio would be desirable but this would depend on the aircraft weight. A good feature is that these wings will always stall at the root end first. This gives good advance warning due to buffeting on the tail from the turbulence streaming from the roots. Lateral control is maintained at the onset of stall because the ailerons are in the undisturbed tip area.

Aspect Ratio 6.3

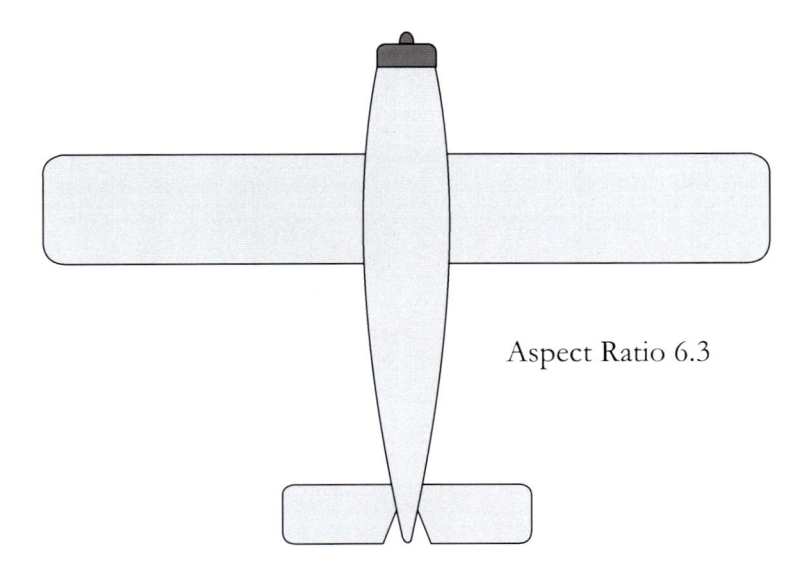

Figure 2.24 - Straight Rectangular Wing

Elliptical

The elliptical wing is the ideal design for low and medium speed use. The downwash it creates is very low and tends to be uniform from tip to root. Due to this it has a low induced drag for the chosen aspect ratio. A further advantage is that the lift coefficient and thus lift is uniform along the span. The shape also allows each section of the wing to be loaded correctly. If the wing stalls it does so uniformly along the span. This would not give much advance warning of a sudden stall and there is a risk of losing lateral control due to the position of the ailerons near the wing tips.

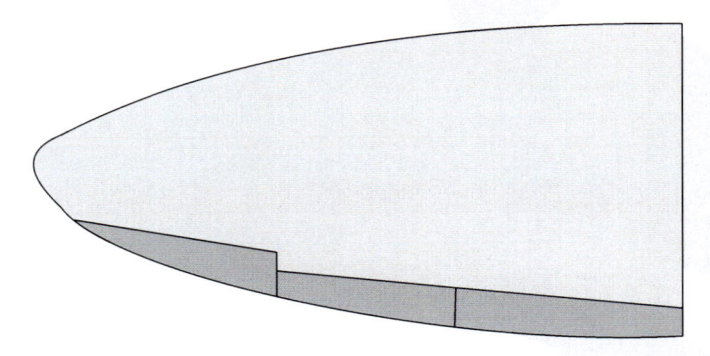

Figure 2.25 - Elliptical Wing

Tapered

Tapered wings are desirable when considering structural weight and stiffness. They do, however, have a problem in that the wing area varies significantly from root to tip. In the extreme case of a pointed tip with wing taper ratio 0 the tip cannot hold the tip vortices so there is little or no downwash at the tip but a strong downwash towards the root indicating that the trailing vortices had moved inboard. The pointed tips would produce little or no lift yet, for the tiny chord length, the lift coefficient there would be relatively high. The pointed tips would appear to be in a state of continual stall. This is a very poor situation.

Lying between the two extremes of a pointed wing and a rectangular wing there are a number of effective compromises. A taper ratio of 0.5 has efficiency close to that of an elliptical wing and that has always been celebrated as the ideal. Unfortunately, it would have the same stall characteristic in that the stall progressed evenly along the span rather than commencing at the root. Lower taper ratios than this would push the onset of stall nearer to the tip and that is not good.

To preserve the aerodynamic efficiency along the span of the mid taper ratio wing they may have to be designed with wash out, reducing camber and thickness/chord ratio towards the tip and higher camber and thickness/chord ratio near the root. This is designed to produce a greater downwash near the root decreasing towards the tip. Hopefully this will encourage the root to stall before the tip.

Tapered wings are swept back on some commercial aircraft. I explain the effect of sweepback nearer the end of this chapter but one problem with this design is that they experience a component of span wise airflow. Because of the span length, this flow can become sluggish and turbulent as it pools at the wing tips. This poses a risk of wing tip stalling. A number of devices are used to combat this including washout, reducing the camber and sweepback angle at the outer part of the wing. The wing may also be provided with a bigger sweepback angle and high camber near the root Boundary layer fences, slots and notched leading edges have also been used to combat this tip stall tendency.

One unusual example of a swept taper wing is that used on the Airbus designs. It is known as the 'supercritical wing'. This has a cross section that is more convex on the underside than on the top. The underside surface reflexes towards concave near the trailing edge. This design spreads the lift more evenly along the chord. A further advantage is that the wing can be structurally much thicker and stronger than conventional wings without producing any more drag.

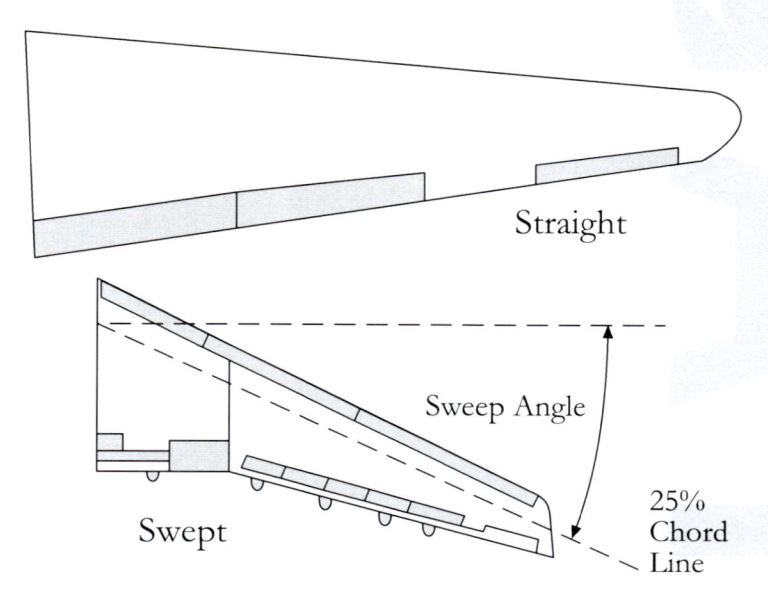

Figure 2.26 - Tapered Wing

Wing Sweep Angle

The sweep angle of a wing is measured as the angle between a line drawn along the span of the wing at the 25% chord length and a line drawn perpendicular to the longitudinal axis of the aircraft.

Swept Back Wing

Again, I will deal more in depth with the effects of sweepback towards the end of this chapter as it has an effect on lift and drag that are topics we have to deal with first. As I have already mentioned, the swept wing has a component of span-wise flow that can stagnate at the tip leading to an airflow separation and tip stall. One problem with a wing tip stall on a swept wing aircraft is that the centre of pressure movement created by it causes the aircraft to pitch nose upwards. This will cause a complete stall if the angle of attack along the span then exceeds the stall angle. A number of devices are used to re-energise the airflow at the wing tip to prevent this separation. These include slots, fences and vortex generators.

I have shown a swept wing in fig 2.27 so that you may see that the air flowing chord-wise is at a lower velocity than the actual aircraft air speed. This is the lift producing flow. It is also the flow that would create a shock wave were it to reach Mach one. Because it travels slower than the aircraft it ensures that aircraft flying close to Mach one will not experience shock wave formation with the resultant sharp increase in drag and turbulence produced by this.

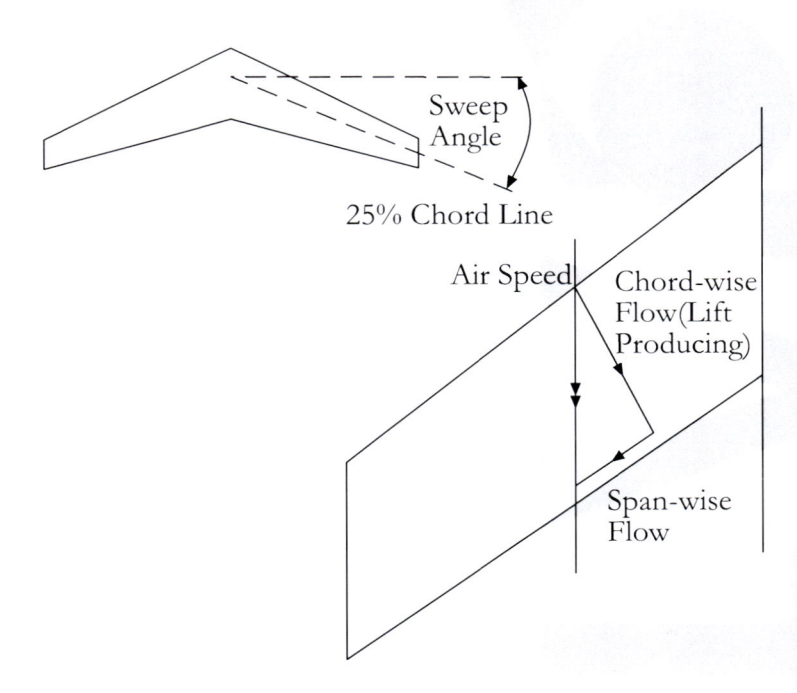

Figure 2.27 - Swept Back Wing

Crescent Wing

The crescent wing has a variable sweep angle along its span together with a reducing wing camber and thickness/chord ratio. The sweep is greatest at the root where the wing is at its thickest. The sweep and section reduces towards the tip where it is almost un-swept. This reduces the span-wise component of airflow in the outboard region, prevents tip stall and minimises the trailing vortex. The crescent wing is a compromise between the swept and un-swept wing. This is now an older design that is little used today.

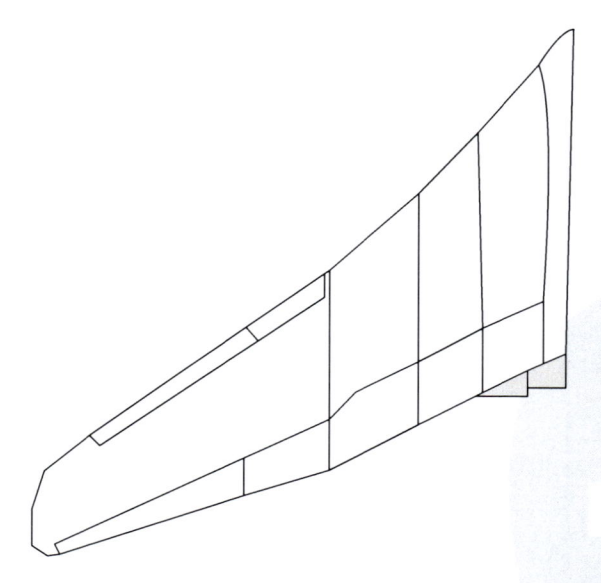

Figure 2.28 - Crescent Wing

Forward Swept Wing

Rarely seen, this experimental type of wing produces a better lift/drag ratio in the lower speed ranges than the swept back wing. One major advantage is that the risk of tip stalling is avoided because the span-wise airflow component acts towards the root. These wings will stall root end first. A drawback is that these wings experience more drag at high airspeeds.

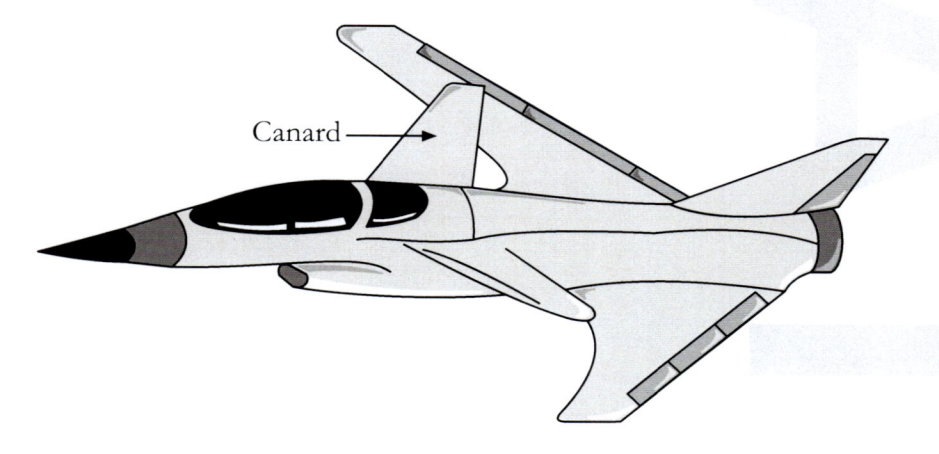

Figure 2.29 - Forward Swept Wing

Delta Wing

There are three configurations of delta wing, the tailless delta, the tailed delta and the slender delta. We will take a very brief look at the first two because it is the slender delta that interests examiners the most. The tailless delta has a very low aspect ratio. They have a very high stalling angle at around 40°, with a marked nose up attitude in the stall. When the trailing edge elevons are used to pitch the nose up at low speed they change the pressure distribution over the wing. The pressure over the wings rises reducing the coefficient of lift at low speed, nose up attitudes. The stall is slow to progress with a gradual reduction in lift. The tailed delta has a tail plane with elevators to control pitch. This design has a lower stalling angle than the tailless type. Both types experience a breakaway of airflow over their wing leading edges at high angles of attack that will form a weak vortex that merely creates turbulence in the aircraft wake.

Slender Delta

The slender delta has a special feature that is not found in any other type of wing. It can produce lift at angles of attack well above those that would produce a stall in other wings. The feature that allows it to do this is called *Vortex Lift*. You have already examined the trailing vortices produced by wings. The slender delta creates a larger, slower moving vortex that emanates from the whole of its wing leading edge and swirls over the top surface of the wing. The wing leading edge is deliberately designed to create the breakaway of air over the wing to ensure that the vortex forms. At high angles of attack the vortex deepens significantly but it is there during the whole of the flight range of angles of attack, albeit with less severity. Unlike ordinary delta wings, the slender delta's vortices remain in place over the wings during flight.

The core of a vortex contains low pressure and its position over each wing enhances the low-pressure region that already exists there, increasing the suction and thus the lift. Vortex lift is the reason why the Concorde could fly slowly at very high nose up angles. This gives rise to the favoured question on slender delta wing aircraft. What happens to the coefficient of lift on a slender delta at low speed at a high angle of attack. The answer is, it increases by up to 30%. Many passengers have reported a series of bouncing motions with a frequency of about 2Hz (two per sec) on take-off. It is the vortices that are producing this. At high speed the centre of gravity position is often adjusted by transferring fuel into the aircraft fin, Concorde did this. The alternative of raising the elevons at the rear of the wing to lift the nose creates another form of drag called *trim drag*.

The slender delta incurs low drag at supersonic speeds because of its low aspect ratio.

Vortex Lift

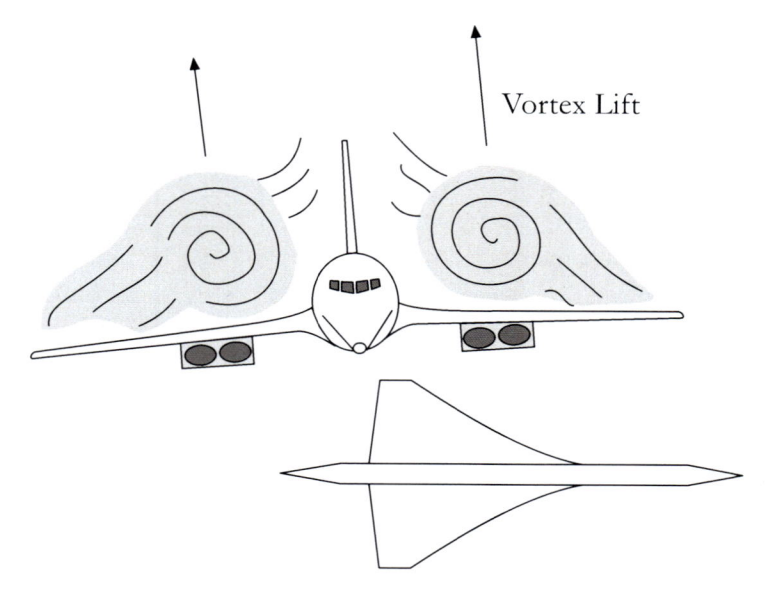

Figure 2.30 - Slender Delta Wing

Polymorphic Wing

Aircraft designed to fly at supersonic speeds often have very poor performance in the low speed ranges, the slender delta wing being the exception. The problem may be overcome by having wings that can translate from a high-speed flight sweep back configuration to a low speed straight wing configuration. The intermediate sweep range is used to suit any specified aspect of flight. This has been achieved on some military aircraft types. Proposals have even been made to extend this to a variable section aerofoil where camber and thickness/chord ratios could be made variable by selection. This remains in purely experimental terms.

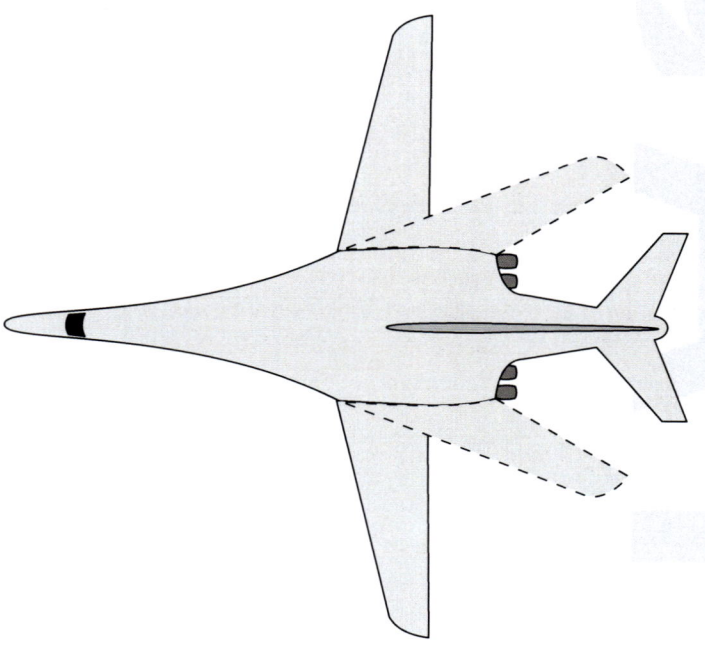

Figure 2.31 - Polymorph

Drag

The drag force acting on an aircraft is defined as sum of all the aerodynamic forces acting parallel to and opposite to the flight path of the aircraft and resist its movement through the air. There are two classes of drag. Profile drag that is created by virtue of the aircraft's shape and surface finish and induced drag that is produced as a side effect of lift.

Profile Drag

Profile drag is not lift dependent and it increases to the square of the airspeed. This means that if the airspeed is doubled, profile drag will increase four times. Triple the airspeed and profile drag will increase nine times and so on. Profile drag may in turn be sub-divided into three sources of drag that together are influenced by shape, surface finish, surface area, air density and air velocity.

Form Drag

Form drag is often referred to as 'boundary layer normal pressure drag'. It is not lift dependent and relies on the shape of the body and the air flowing over it. If you can imagine a flat plate presented to the airflow at an angle of 90°, as shown in figure 2.1a. You would see that the air passing around it separates and re-circulates at its rear creating a low-pressure region. Taken in conjunction with the atmospheric pressure acting on the front of the plate a pressure difference will exist that will act on the frontal area of the plate to create a force that will either try to push the plate rearwards or, resist its passage forwards.

If the angle of the flat plate to the airflow were reduced to 45° you would still see that the airflow passing it would separate and re-circulate behind it but to a lesser degree. The pressure difference created now acts on a much smaller frontal area and the rearward force is thus less than for a plate at 90°. Finally, if the plate were presented to the airflow at zero degrees there would be no separation and the drag created would stem from the friction arising from the shearing of the layers of air in the boundary layer.

When we examine other shapes such as those shown in figure 2.1b, 2.1c and 2.1d we can observe that the separation point where the boundary layer leaves the surface and re-circulates occurs successively further back on the body. To sum up, the further forward the boundary layer separation points are with subsequent increased turbulence in the wake, the higher the form drag will be. Form drag is dependent upon the separation point that is in turn influenced by the transition point, adverse pressure gradients and streamlining. A smooth, streamlined shape is therefore the objective in reducing this type of drag.

Skin Friction

Skin friction or surface friction drag as we already know stems from the friction in the boundary layer that is created by the shearing of air molecules in the sub-layers. The magnitude of this type of friction is dependent upon the surface area and finish, the position of the transition point, air density and air velocity. Where the boundary layer becomes turbulent after the transition point the skin friction will increase. Skin friction can be reduced in value by producing a highly polished surface finish. We can also say that these factors in many ways influence form drag so that places them under the single title of profile drag.

Interference Drag

Interference drag stems from the junctions that are formed by the assembled components of the aircraft. Examples of these are the wing to fuselage joints, the fin and tail plane joints and the engine nacelle to wing joints. The mixing of airflows and the modification to the boundary layers in these regions creates additional turbulence in the wake that is added to that already created by the form drag. It is due to this that interference drag is classified together with form drag and skin friction under the heading of profile drag. Interference drag may be reduced by the use of smooth, curved fairings and fillets at the intersection joints.

Calculation of Profile Drag

The calculation of profile drag requires that we have some measure of the shape and surface condition of the particular body in question. This means we need a coefficient which is a value used to modify the general calculation for drag so that it gives a true result for the body we are examining. A simpler explanation for how this coefficient is arrived at is to consider a worst situation, the flat plate performance for example, and call that a coefficient of 1. Now figures can be produced experimentally to reflect the performance of different shapes at different angles of attack. This gives us coefficients for streamlined shapes that are typically in the region of 0.015 to 0.04. Shows you the effect of streamlining and polishing!

Now we need a few other figures. The plan form area of the body, the density of the air and the velocity of the air-flow. Once we have these we can use the following formula:

$$\text{Profile Drag} = C_{dp} \times \tfrac{1}{2}\rho v^2 \times S$$

Where:
  C_{dp} is the coefficient of profile drag (No units reqd)
  ρ is the air density in kg/m³
  v is the air velocity in m/s
  S is the plan form surface area in m²

The units shown are SI units and will give a drag force value in Newton's. You could of course use imperial units and gain a drag force value in lbs force.

If you examine the formula you will see that what we have done is multiply the dynamic pressure $½\rho v^2$ by the surface area and then modified the result by multiplying it by the coefficient of drag for the body. You will also find that if you then doubled the air velocity value the profile drag value would increase by four times.

Induced Drag

Induced drag is lift dependent and unlike profile drag it reduces to the square of the air velocity. Double the airspeed and the induced drag will be quartered. At low airspeeds, particularly during climbing, the induced drag can account for up to 75% of the total drag on the aircraft. At high speeds it will account for about 10% of the total drag. We have discussed how this type of drag is created. Do you recall the formation of the trailing vortices and how they swirl in behind the wing to push the airflow leaving the wings down creating a downwash. Recall also that it is the downwash that alters the direction of the relative airflow passing over the wings so that an increase in downwash reduces the wing angle of attack and tilts the resultant lift vector to the rear. This creates a rearward acting component of force we call induced drag.

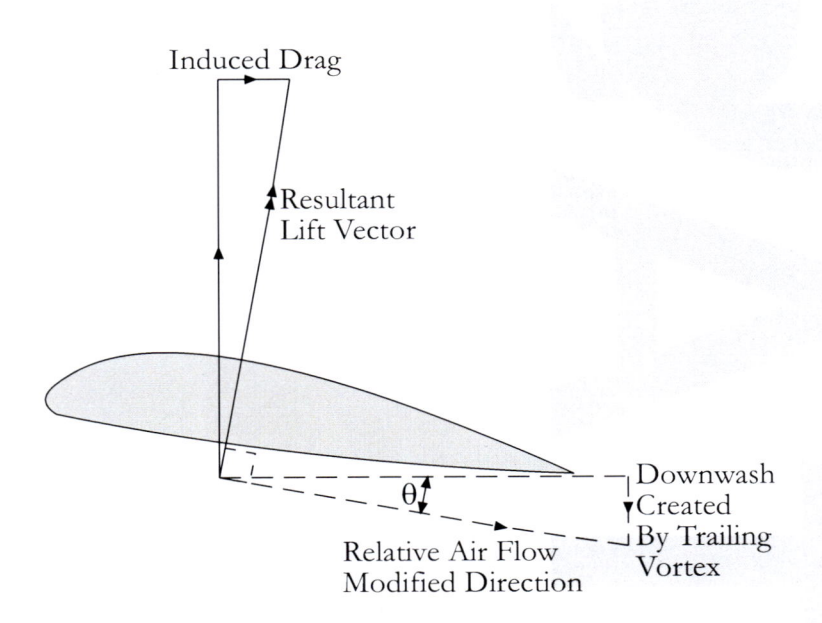

Figure 2.32 - How Lift Induced Drag Occurs

Induced drag is created as a side effect of lift. If the lift produced by the wings increases, the pressure difference between the upper and lower wing surfaces increases and the air spilling over the tips increases. This increases the intensity of the trailing vortices and as a result increases the downwash. The resultant lift vector tilts further back and the induced drag component gets bigger.

High aspect ratio wings tend to have reduced trailing vortices and thus lower lift-induced drag. This is due to the smaller 'end or tip effect'. The wing tip chord is small therefore the amount of air spill is less than if it were longer. The coefficient of drag and thus drag itself are reduced by the use of high aspect ratio wings.

Induced drag will be highest when the lift force is at its greatest value. This usually occurs at low airspeeds when angles of attack are high and often the flaps are extended. This would be typically take-off and landing. At higher airspeeds, such as cruising speeds, the angle of attack of the wings will be low, around 3 to 4° and total lift force will be at the minimum required to maintain level flight, thus induced drag will be low.

When aircraft are increasing in altitude, the air density will be reducing and this will reduce lift. This necessitates an increase in the angle of attack to restore lift and this will increase downwash and thus the induced drag. The profile drag will of course reduce significantly with an increase in altitude so the total drag will reduce. If the aircraft weight is increased, more lift will be required. This again requires that the angle of attack be increased to increase lift. The induced drag will then increase. The formation of ice on an aircraft in flight is an example of one way the weight might increase.

The trailing or wake vortices that produce induced drag are more intense behind heavy aircraft. This is simply because these aircraft have to produce more lift to stay aloft.

Induced drag can be reduced by the use of 'winglets' at the wing tips. These have the effect of lifting the trailing vortices so that their influence on downwash is reduced. The joint between the winglet and the wing tip also produces a tight vortex that rotates counter to the trailing vortex and modifies it. In addition to this the winglets are angled to give a forward component of 'lift', more accurately described as thrust. This acts to counter the backward component of induced drag. Designers of aircraft using winglets claim that these devices increase range and reduce fuel consumption because of their effect on reducing induced drag.

As an alternative to the winglets, wingtip fences may be used. These project above or below the wingtips, sometimes both. They serve to reduce the spill of air from under the wings around the tips and into the low-pressure regions above the wings. This again reduces the trailing vortices.

Winglet

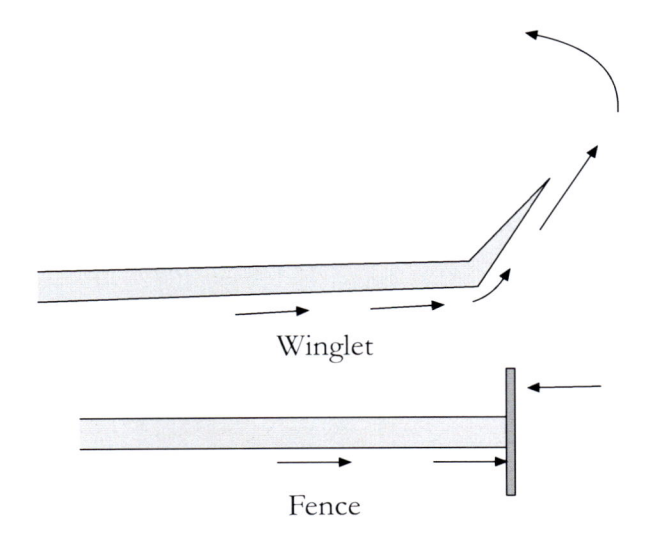

Fence

Figure 2.33 - Devices Used to Modify Trailing Vortices

To sum up, the main factors that influence induced drag are: Plan form, aspect ratio, lift and weight and airspeed.

Calculation for Induced Drag

Once again we need a coefficient, this time for induced drag. This is calculated as follows using as an ideal example the formula for elliptical plan form wings. For other shaped wings we would have to multiply the bottom line by a correction factor e.

$$C_{di} = \frac{C_L^2}{\pi A}$$

Where: C_{di} is the coefficient of induced drag (no units)
         C_L is the coefficient of lift (no units)
         π is 3.147 (no units)
         A is aspect ratio (no units)

Notice that from the above equation that if you double the aspect ratio you would halve the coefficient and thus halve the induced drag.

The induced drag will be:

$$C_{di} \times \tfrac{1}{2}\rho v^2 \times S$$

Do not get confused here, the induced drag does not increase to the square of the airspeed it decreases to the square of speed. Note that the coefficient of lift was used to establish the coefficient of induced drag and at higher airspeeds this coefficient will reduce and thus reduce the induced drag coefficient.

I have included the calculations for induced drag to counter any imaginative examiners. Normally they will stick with the overall formula for total drag.

Total Drag

To calculate the total drag force we need to have a coefficient for total drag. This is derived by the addition of the coefficients for profile drag C_{dp} and induced drag C_{di} giving a total drag coefficient C_D. You will need to be able to transpose the total drag formula around to show it for both drag and for the coefficient.

$$\text{Drag} = C_D \times \tfrac{1}{2}\rho v^2 \times S$$

$$C_D = \frac{\text{Drag}}{\tfrac{1}{2}\rho v^2 \times S}$$

Where:
- C_D is the coefficient of total drag
- ρ is the density of the air
- v is the air speed
- S is the plan form area

Note: $\tfrac{1}{2}\rho v^2$ is the formula for dynamic air pressure. This represents the kinetic energy of the airflow and it is this value that is proportional to the square of the air velocity.

Thrust & Weight

Thrust

In the previous section you were introduced to the drag forces that act to resist the forward motion of an aircraft in flight. We will be returning to that topic later on. At this point we need to examine the force that opposes these drag forces in order that the aircraft may be flown at constant speed in level flight, accelerated or climbed.

To achieve a state of equilibrium where the aircraft can maintain a constant velocity in level flight the thrust force must exactly balance the drag forces. If the aircraft is to be accelerated to a higher velocity then the thrust force must exceed the drag forces for the period of the acceleration.

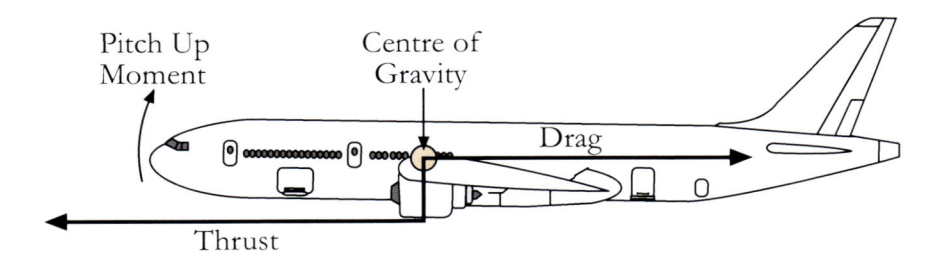

Figure 2.34 – Thrust

In climbing flight, the thrust force must be increased in order to exceed the sum of the drag forces and the component of the aircraft weight that will be acting rearwards as a result of the climbing angle. In the latter case, the thrust will be supporting part of the aircraft weight during a climb. Due to the climbing angle, the lift produced by the wings will always be less than the aircraft weight during a climb.

We will be dealing with the arrangement of the forces acting on an aircraft in the next chapter. For now, you should merely appreciate that a large part of the level flight and climbing performance of an aircraft will depend on the amount of thrust available from its engines. The most fuel efficient and quietest engine configuration in use today is the turbo-fan engine. These engines rely on reaction thrust in accordance with Newton's second and third laws. The jet efflux speed is matched to be close to the aircraft speed for maximum propulsive efficiency. These engines also depend on air density for their thrust performance. The available thrust reduces significantly as altitude increases. Fortunately, the drag forces acting on the aircraft also reduce in the same proportion as altitude increases.

Weight

Weight is another force that acts on an aircraft. In level flight the weight is the product of the aircraft mass and the acceleration due to gravity. Upward acceleration will add to this and downward acceleration will subtract from it. Again, the way it interacts with the other three flight forces, lift, thrust and drag, will be discussed in more depth in the next chapter.

The weight will always act vertically downwards through the aircraft centre of gravity regardless of the attitude of the aircraft.

To maintain equilibrium in level flight at constant altitude the weight must be balanced by the equal and opposite force of lift. If the lift exceeds the aircraft weight, the aircraft will gain altitude. If the lift is less than the aircraft weight, the aircraft will lose altitude.

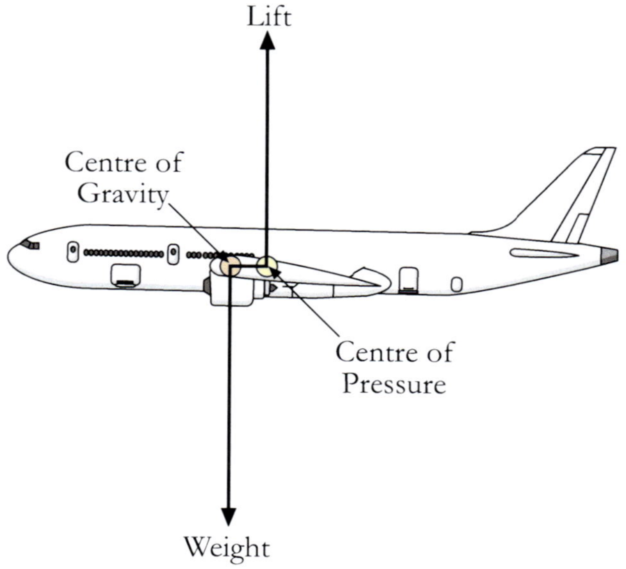

Figure 2.35 - Weight

When an aircraft is being manoeuvred during a turn it is essential that the lift will balance the weight. During a climbing manoeuvre, the lift will be less than the weight of the aircraft so as we have mentioned earlier, the thrust must be increased in order to make up the difference. In a vertical climb, the weight would need to be balanced wholly by thrust.

Weight is a critical value as the aircraft cannot lift off or maintain altitude unless it can produce sufficient lift to oppose and occasionally exceed its weight. Lift is significantly affected by air density so the maximum allowable weight of an aircraft will be determined by a number of factors such as air temperature and air pressure, both affecting air density.

Even if the aircraft can find the necessary lift, there must be a limit to the weight to prevent overstressing the aircraft wings.

The distribution of weight and thus the position of an aircraft centre of gravity are of primary importance. If the centre of gravity were to be too far forward the aircraft would tend to fly nose down. If the position were too far to the rear then the tendency would be reversed to a nose up attitude. In either case, valuable aircraft control movement would have to be employed to correct the situation. This would then deny some of the available control facility for use during normal flight. Movement of the centre of gravity does have an effect on stability. In the extreme case where the centre of gravity were too far displaced either way the aircraft may not have the necessary control range of movement to correct the situation and the result could be a disaster.

The weight of an aircraft in smooth and level flight is the result of the earth's gravitational pull acting on the mass of the aircraft. In the example we have just given, the aircraft will be experiencing 1G. Imagine that the aircraft were then to meet turbulence or, be put into manoeuvres where it may experience a degree of vertical acceleration either up or down. Weight is a force that is derived from the product of mass and the acceleration due to gravity. Any upward acceleration will therefore increase the weight of the aircraft and downward acceleration will decrease it. We will discuss this effect in more detail in chapter three but let it suffice to remember that these variations in aircraft weight have to be matched by lift forces and this will put demands on the stress and strain limits of the aircraft structure.

Aerodynamic Resultant

Lift is the sum of all the components of the aerodynamic forces that act perpendicular to the flight path through the centre of pressure. Drag is the sum of all the components of the aerodynamic forces that act parallel to and opposite to the direction of flight and oppose the aircraft movement.

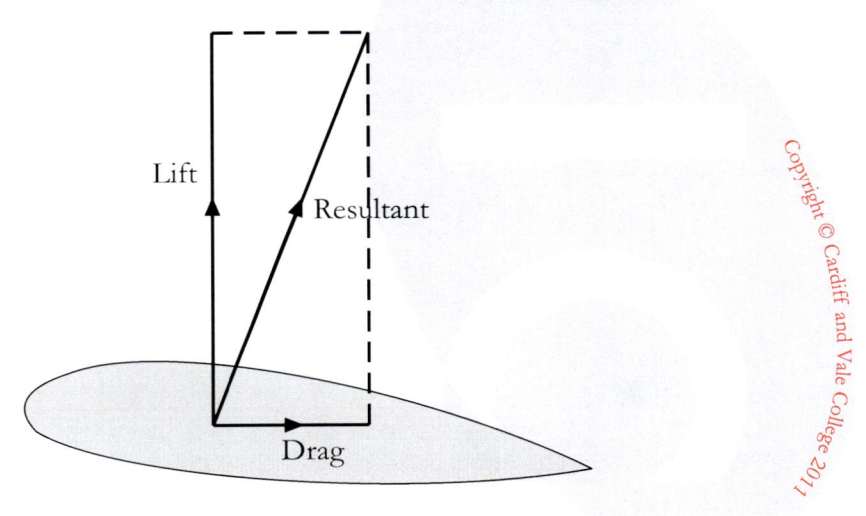

Figure 2.36 - Resultant of the Lift and Drag Forces

We can resolve these two forces to produce a resultant force that will replace both. This resultant force inclines to the rear to a degree that is dependent on the lift/drag ratio. If the aerofoil design can be improved to reduce drag and to increase lift, the resultant force will be less inclined away from the vertical. It is the vertical component of force that opposes the weight so it is desirable that the resultant is as close to the vertical as is possible.

Generation of Lift & Drag

Lift

Lift arises from the difference between the pressures on the upper and lower surfaces of the aerofoil. The sum of the components of lift acting on the aerofoil can be resolved into a single force that acts at right angles to the direction of the resultant airflow and passes through a single point on the aerofoil chord line called the centre of pressure (CP).

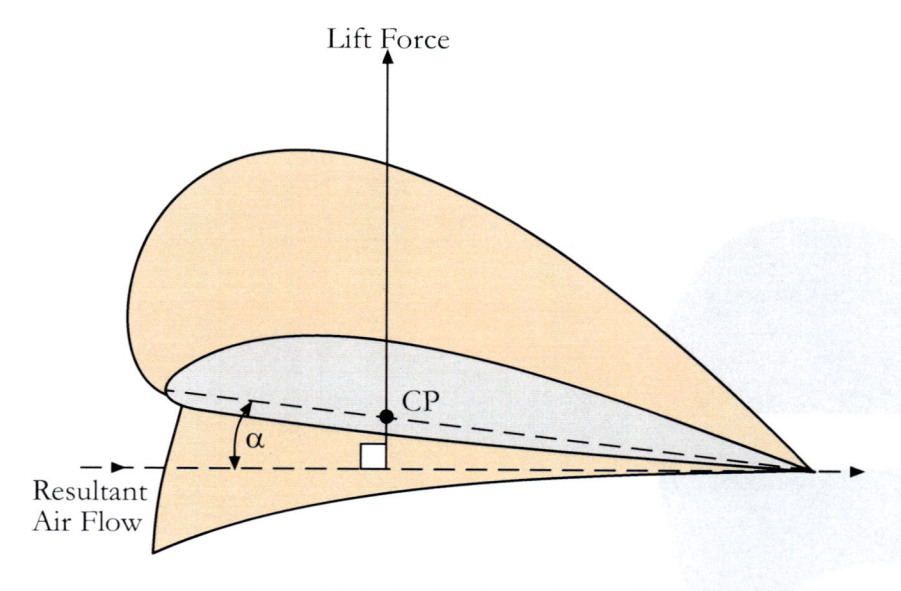

Figure 2.37 - Angle of Attack and CP

The angle made between the aerofoil chord line and the path of the resultant airflow is the angle of attack (α). When this angle is a negative value for example at -3° or less, no lift will be produced as the low-pressure regions above and below the aerofoil will be of equal value and effect.

At 0° angle of attack, lift will be produced providing the aerofoil is asymmetric, that is, it has a curved upper surface with a flat or very low curvature underside. In this case a difference will exist between the pressure reductions on the upper and lower surfaces. The centre of pressure will be well to the rear of the chord line at this stage. If the aerofoil is symmetrical, having equal curvature top and bottom then no lift can be produced at 0°. This latter aerofoil would depend on having a positive angle of attack to produce any lift.

At +3° angle of attack lift increases and the centre of pressure will move forwards to between 30 and 40% of the chord length from the leading edge. The lift is still arising from the difference between the pressure reductions on the upper and lower surfaces, much lower pressure at the top than at the bottom.

At +7° angle of attack lift increases further and higher pressure starts to appear on the under surface. The centre of pressure will have moved forward again to between 20 and 30% of the chord length.

CHAPTER TWO
AERODYNAMICS

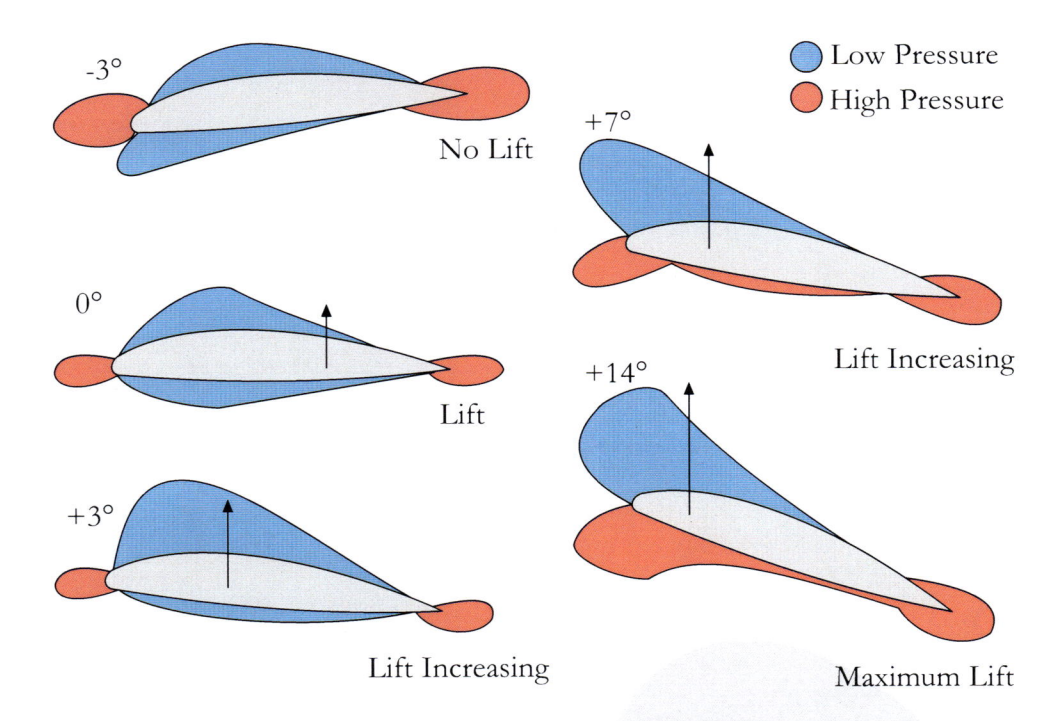

Figure 2.38 - Aerofoil Pressure Distribution

At 14° angle of attack the high pressure under the aerofoil will be contributing to a larger part of the lift, the low pressure region over the top will have deepened and moved forward. The centre of pressure will be at its fully forward position at about twenty percent of the chord length.

At the stall angle of 15° or over, the low pressure over the top section will suddenly reduce and the little lift remaining will be due mainly to the high pressure existing under the aerofoil. The centre of pressure will move quickly rearwards.

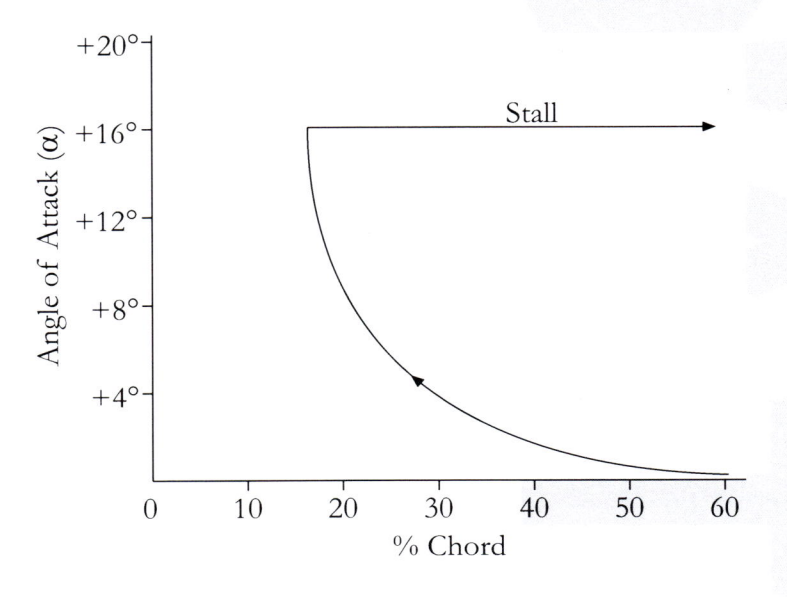

Figure 2.39 - Movement of Centre of Pressure

The lift force may be calculated and this is similar to the method we used to calculate total drag. We need to establish a coefficient for lift (C_L). This figure may be established from a set of tables appertaining to the aerofoil section at different angles of attack, for example NACA or RAF tables. We then need the wing plan form area, the air density and the air velocity. The calculation is as follows.

$$\text{Lift} = C_L \times \tfrac{1}{2}\rho v^2 \times S$$

Where:
    C_L is the coefficient of lift
    ρ is the air density
    v is the air velocity
    S is the wing plan form area

Note: $\tfrac{1}{2}\rho v^2$ is the formula for dynamic air pressure. This represents the kinetic energy of the airflow and it is this value that is proportional to the square of the airspeed.

Examine the above formula and see if you can understand that if the air velocity were to double the lift would increase by four times. From this you can construe that the lift is proportional to the square of the air velocity. Looking further you will see that it is the dynamic pressure that gives it this relationship so you can also say the dynamic pressure is proportional to the square of air velocity. Now look further, pressure × area = force so you can now say that the aerodynamic force is proportional to the square of the air velocity. That is why aircraft controls become hard to move at higher airspeeds if you do not have a power assist system.

Be prepared to see rearrangements of this formula to see if you know what is in it. Also be prepared to transpose it to find the coefficient of lift as follows:

$$\text{Coefficient of Lift } C_L = \frac{\text{Lift}}{\tfrac{1}{2}\rho v^2 S}$$

Coefficient of Lift

The coefficient of lift is a figure that is used to convert the aerodynamic force of lift from its theoretical value to that applicable to a particular aerofoil design at a given angle of attack. We already know that lift increases with an increasing angle of attack up to the stall angle of about 15°. The lift coefficient increases in line with the lift force so it is a useful figure when used in isolation, as it is a measure of the relative value of lift.

The ideal angle of attack for a wing is between three and four degrees but this does not give you the maximum lift it gives you the best lift/drag ratio. At this angle, lift can be 25 times the value of drag. As the angle of attack increases, lift will rise but drag will rise at a faster rate so at high angles of attack lift may only be 6 times greater than drag.

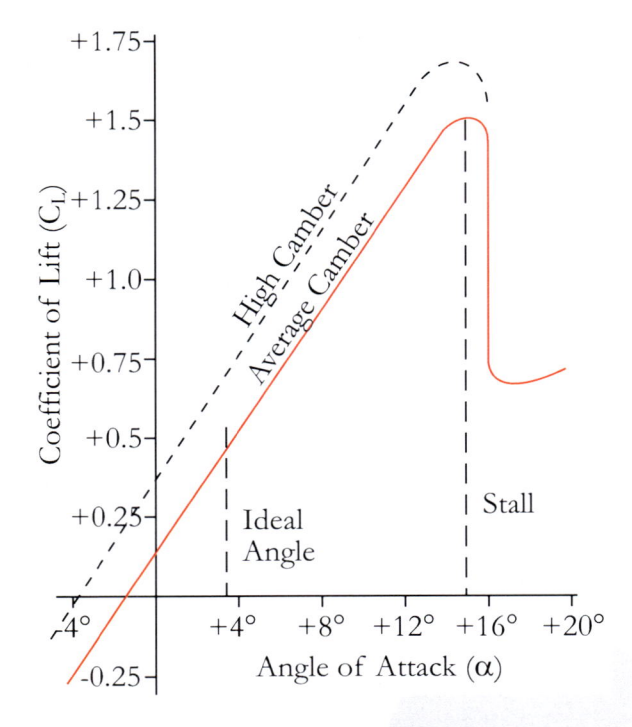

Figure 2.40 - Lift Curve for a Typical Aerofoil

The value of the lift coefficient will be greater for highly cambered aerofoils up to a limit. We have also mentioned previously the vortex lift experienced by slender delta wing aircraft. This phenomenon increases the coefficient of lift for those aircraft particularly at low airspeeds and that is unusual for a high speed swept wing aircraft.

Drag

We have already examined how drag is generated in both its profile form and its lift induced form and we have learnt how to calculate the total drag.

I would like to remind you how lift induced drag occurs. The lift force acts at right angles to the path of the relative airflow. The downwash off the trailing edges of the wings is influenced by the effect of the wing tip trailing vortices. These are created by air spilling out from under the wings over the wing tips and then swirling in towards the low- pressure regions over the wings. The downwash alters the path of the relative airflow reducing the angle of attack. As the lift force is always at right angles to the path of the relative airflow, its line of action is tilted rearwards. This creates the rearward component of force we call induced drag.

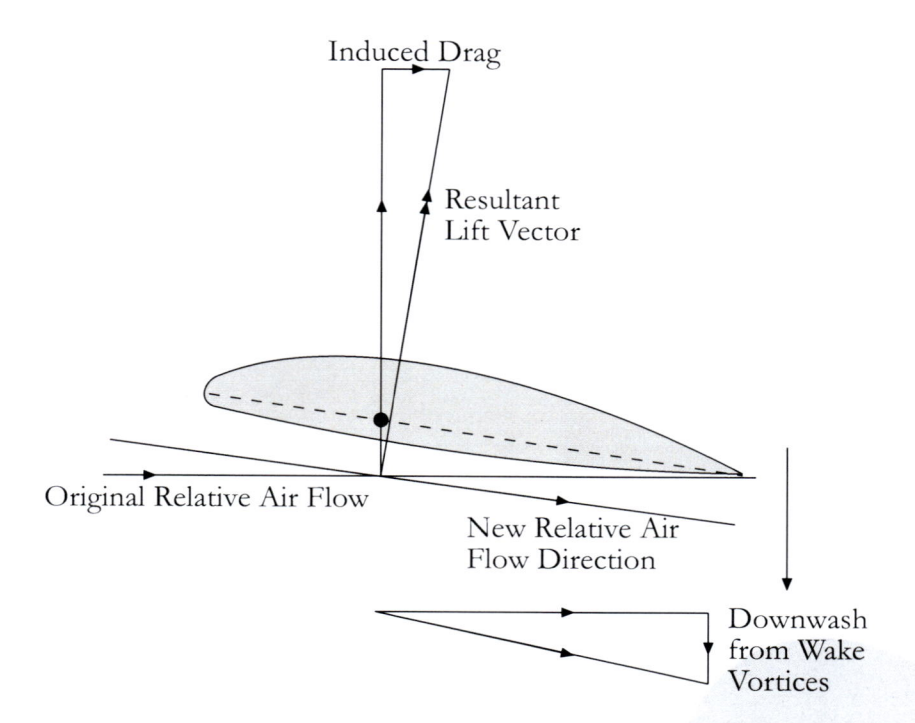

Figure 2.41 - Lift Induced Drag

If the lift of the wings is increased the wing tip vortices intensify, the downwash angle increases, angle of attack reduces and the induced drag increases. Any increase in the angle of attack made to increase lift will increase the induced drag. Total drag is the sum of induced drag and profile drag. When this total is shown together with the lift force as a vector diagram we can see how it produces the aerodynamic resultant force.

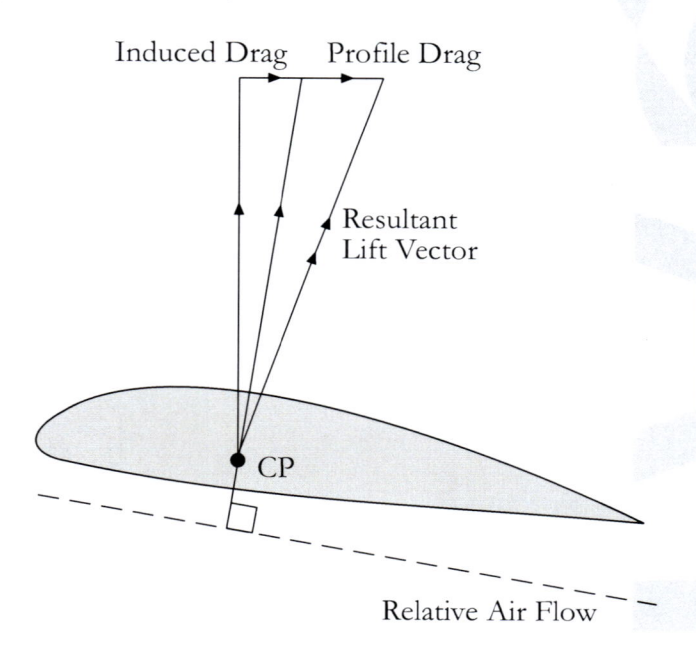

Figure 2.42 - Total Drag

As the angle of attack of an aerofoil is increased both the induced and profile drag values increase. If you recall, when we looked at the lift curve we used a coefficient. We do the same with drag. The coefficient of drag is a convenient method of displaying values that relate to drag.

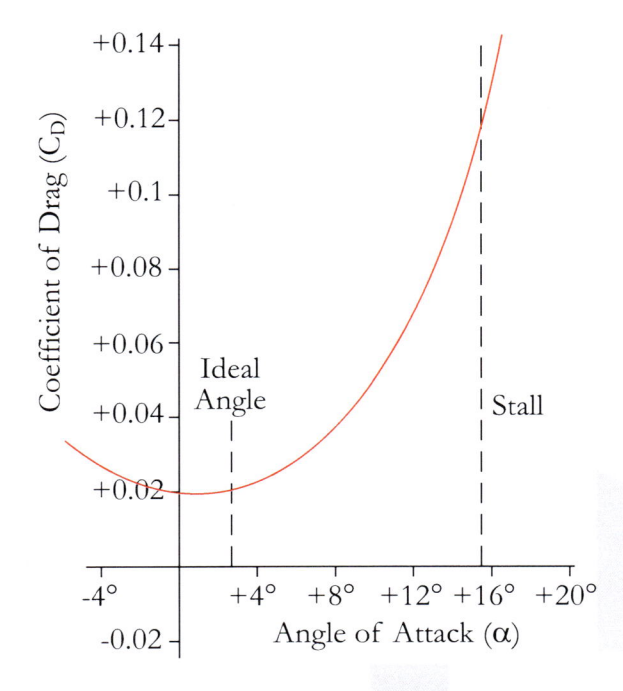

Figure 2.43 - Drag Curve for a Typical Aerofoil

Again the ideal angle of attack is between three and four degrees. Drag will increase up to 15° angle of attack; beyond that the aerofoil stalls and drag increases dramatically.

To examine the relationship that drag has with aircraft speed we need to treat profile drag and induced drag separately.

Profile drag is proportional to the square of airspeed. If the airspeed doubles the profile drag increases four-fold. Triple the airspeed and profile drag increases nine-fold.

Induced drag is inversely proportional to the square of airspeed. If the airspeed doubles the induced drag is quartered. If the airspeed triples, the induced drag is reduced to a ninth of its original value.

Total Drag

When we examine the sum of profile drag and induced drag for increases in airspeed we should find that there is a particular airspeed where the total drag acting on the aircraft will be at a minimum value. At low speeds during take-off and approach and landing, the profile drag will be very low but the induced drag will be very high, as a lot of lift will be required to maintain flight. As airspeed is increased, the profile drag increases and the induced drag decreases.

If we sum up the total drag acting on the aircraft you will see that the profile drag curve crosses the induced drag curve at a particular airspeed that is referred to as the ***Minimum Drag Speed***. Aircraft usually try to select a speed just above this so that if they encounter a gust of wind that tends to slow the aircraft, the drag reduces and the aircraft immediately regains its set airspeed. This possibly gave rise to the saying 'keeping ahead of the drag curve'.

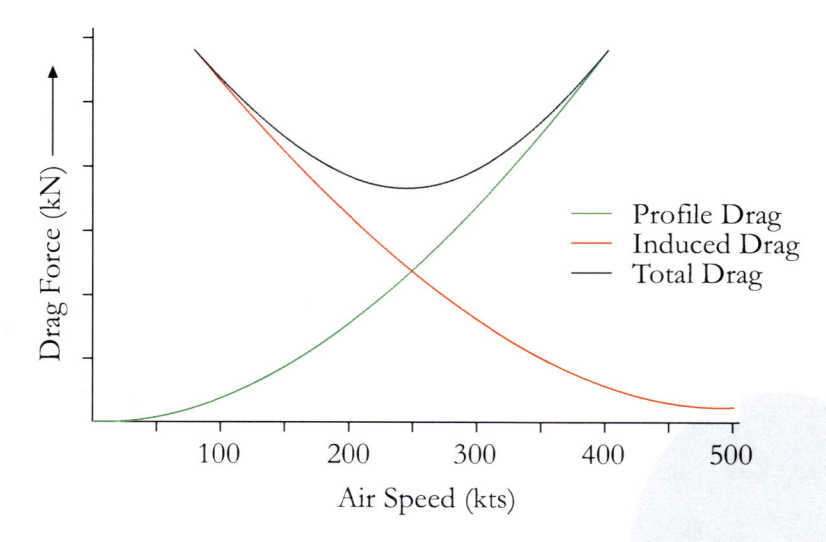

Figure 2.44 - Total Drag Curve

Coefficients of Lift & Drag

Earlier in this chapter we examined the lift to drag ratios at varying angles of attack. This implied the use of values representing the actual forces. It is useful to use the coefficients of lift and drag as these can be plotted with more accuracy. We simply examine how many time the drag coefficient divides into the lift coefficient for each angle of attack.

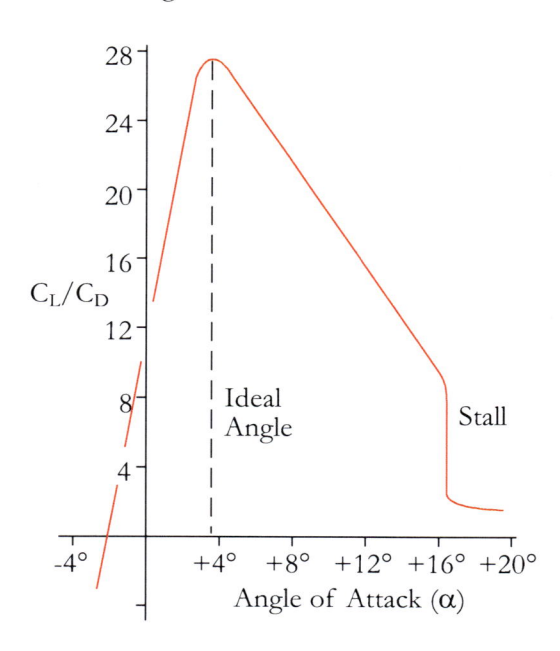

Figure 2.45 - Lift/Drag Ratios using Coefficients C_L/C_D

Polar Curve

A polar is a term representing a graph of C_L against C_D for various airspeeds. They are used to display the drag characteristics of an aircraft at each speed. They show how the coefficient of drag will increase in relation to increasing lift coefficients. The drag experienced by high-speed aircraft is far greater than those designed for lower speeds. High transonic and supersonic aircraft can experience three times more drag. This accounts for the higher thrust requirements and subsequent fuel consumptions of aircraft similar to the late Concorde.

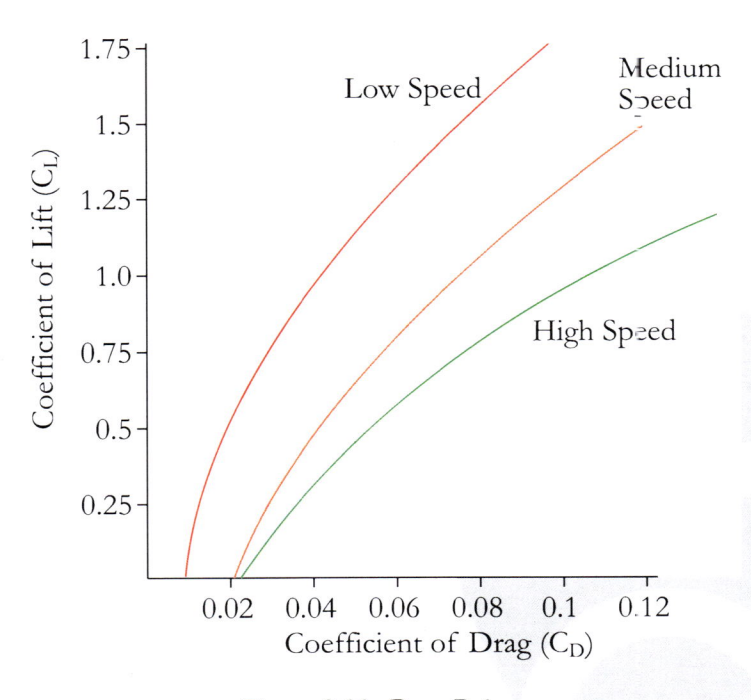

Figure 2.46 - Drag Polars

Stall

An aerofoil is considered to have stalled when it reaches an angle of attack where the airflow can no longer follow its contours and separates completely from the upper surface causing a significant reduction in the coefficient of lift. Airflow re-circulation behind the separation point will create turbulent eddies that increase drag at the onset of a stall.

To understand why the airflow should separate in this way you need to examine the pressure gradient that exists over the upper surface of an aerofoil. The free-stream airflow accelerates from the leading edge up to the point of maximum curvature creating a proportional fall in pressure. As the airflow passes back from this point heading towards the trailing edge, the pressure will be rising again until it finally regains the free-stream pressure. The pressure gradient from the point of maximum curvature to the trailing edge is referred to as an *adverse pressure gradient*. The air passing along it will decelerate against the rising pressure. In normal circumstances there is enough kinetic energy in the airflow to enable it to reach the trailing edge. As the angle of attack is increased, the adverse pressure gradient also increases slowing the airflow further and eventually causing it to stop and reverse flow. At a given angle of attack, typically around 15° the airflow will separate completely from the upper surface in a re-circulatory flow and stall will have occurred.

The lift/drag ratio of an aerofoil is at an optimum value at around 4° angle of attack. As the angle of attack increases, drag increases faster than the lift. As the stall is approached, the re-circulatory flow that commenced at the trailing edge creeps forward over the aerofoil top surface. Just prior to reaching the stall angle the coefficient of lift will be at its maximum value. As the stall angle is reached the flow suddenly separates over the whole top surface.

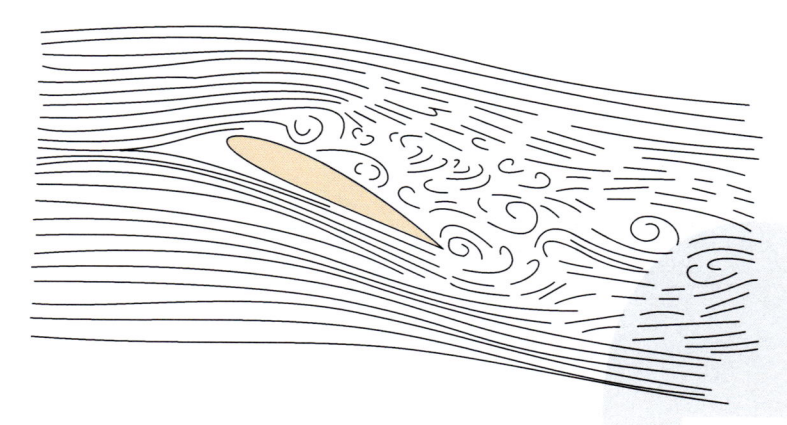

Figure 2.47 - Stalling of an Aerofoil

It is important for you to remember that an aerofoil stalls **at a given** angle not at a given speed. The airspeed that coincides with the aerofoil reaching the stall angle will vary according to circumstances.

Imagine for a moment that an aircraft is flying at a constant airspeed and altitude. The pilot then gradually reduces the engine power. Drag will now exceed thrust and the aircraft will begin to slow down. Lift will reduce and the aircraft will lose height unless the angle of attack is increased to regain lift. This will increase the induced drag and the aircraft will slow further necessitating yet another increase in angle of attack. If this situation is allowed to persist the wing will eventually reach the stalling angle of attack of 15°.

Now imagine that another aircraft of exactly the same type is flying alongside our aircraft doing precisely the same thing. The difference is that the second aircraft is fully laden whereas our aircraft was empty. The heavier aircraft needs more lift than ours to keep it aloft. The result will be that the heavier aircraft will reach the stalling angle of attack of 15° before we do. This means it will experience stall at an air speed above the speed where we would reach stall. We may learn from this that an increase in aircraft weight will increase value of the minimum airspeed at which the aircraft will stall.

Imagine that our two similar type aircraft are now at the same weight. The difference now is that we are flying at low altitude whilst our partner is flying at high altitude. We both commence to lose air speed. Lift is dependent on air density so the higher altitude aircraft will need to maintain a higher angle of attack than us to produce enough lift. Guess what? The higher altitude aircraft will stall first. Again we learn from this. As altitude increases, the speed at which an aircraft will stall increases.

Air density is altered by conditions other than altitude of course. An increase in ambient air temperature will reduce air density and thus increase an aircraft's stalling speed. A drop in ambient air pressure has a similar effect.

Remember, an aerofoil stalls at a given angle not speed. The air speed at which an aircraft wing will reach this angle alters according to the aircraft's weight and/or the density of the air.

Stall does not occur simultaneously over all parts of a wing. Aircraft wings are normally designed to ensure that a stall will commence at the wing roots first. This may be ensured by, either introducing washout at the wing tips to reduce the angle of attack there or reducing the aerofoil section and camber at the tips. Tip stalling is a hazardous occurrence as the sudden loss of lift can cause a wing to drop putting the aircraft into a spin. Lateral control would also be badly affected due to the position of the aileron at the wing tip.

Severe icing on an aircraft will not only disrupt the aerofoil performance and reduce lift but it will also increase the aircraft's weight. This would result in experiencing stall at a totally unpredictable air speed.

Before we move on I must point out to you that you may hear of two types of stall, low speed stall and high-speed stall. We have been examining the low speed stall. The high-speed stall is created by a completely different method. It results from an aircraft experiencing Mach one airflow over the point of maximum curvature on its wings. This creates shock waves that destroy lift. The high-speed stall is sometimes referred to as the shock stall for that reason. The aircraft speed at which it may occur is known as the critical Mach No. of the aircraft. The topic of high-speed flight is covered in more depth in module 11. I will, however, include for you a section on the effects of wing sweep.

The Effects of Wing Sweep

The wings of high-speed aircraft are swept back to reduce drag at the higher speeds. The high-speed low drag advantages are paid for by a degraded performance in the lower speed ranges.

Effect of Sweep Back on Lift

A swept wing has a lower maximum coefficient of lift that a straight wing. The reduction can be as much as 30% for wing sweep angles around 45°. The reasons for this may be found in the span-wise airflow that is created by swept wings and the airflow separation that occurs on the top of the wing towards the wing tips.

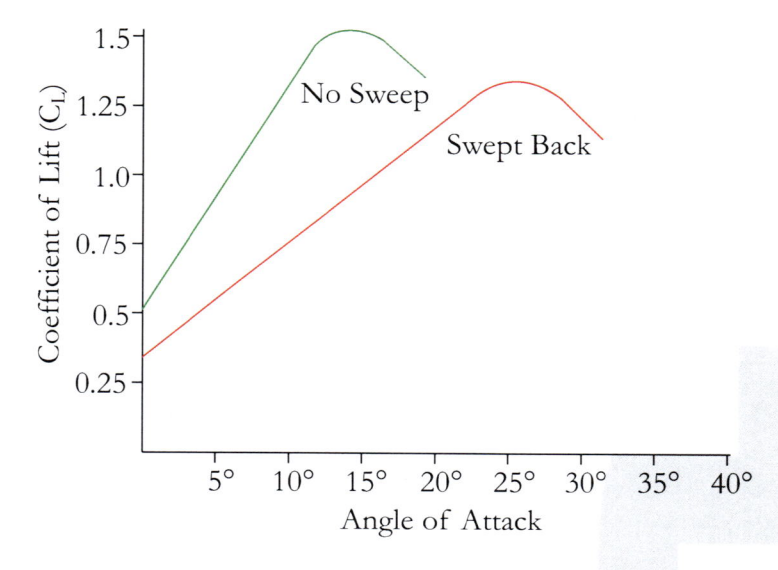

Figure 2.48 - Effect of Sweep Back on Lift

When the free-stream airflow (V) meets a swept leading edge it divides into two components of flow direction and velocity. The first (V1) passes across the wing parallel to the chord line. The second (V2) moves along the span of the wing towards the tip. If the sweep angle is increased, the chord wise flow velocity reduces and the span-wise flow velocity increases. At all angles of sweep, the lift producing chord-wise airflow velocity will be less than the free-stream air velocity.

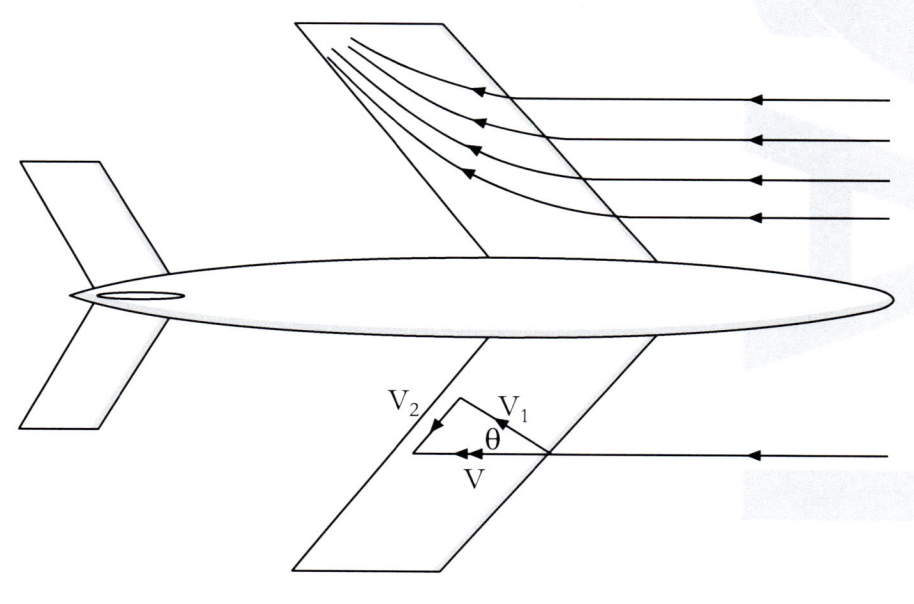

Figure 2.49 - Airflow Over a Swept Wing

> Chord-wise flow Vl = V × Cosine of Wing Sweep Angle(θ)

This means that the lift producing capacity, or coefficient of lift, is proportional to the cosine of the wing sweep angle. The bigger the wing sweep angle, the lower the coefficient of lift will be.

Swept wing aircraft achieve their maximum coefficient of lift at a higher angle of attack than straight wings. Swept wings achieve it around 28° compared to 15° for straight wings. This accounts for the higher stalling angles attributed to swept wings. Aircraft having low aspect ratio swept wings cannot utilise the angles of attack that give the highest coefficients of lift for landing because of the very high nose up attitudes this would necessitate. They have handling problems and anyway, the tail would scrape along the ground! So, they have to employ higher landing speeds. Do remember though that the slender delta capitalises on vortex lift so it is an exception.

Effect of Sweep Back on Drag

The swept wing experiences higher induced drag than a straight wing at higher angles of attack. One obvious reason why the drag is increased is that the low coefficient of lift means that a higher angle of attack has to be set to give the required lift. This increases downwash and induced drag. When the angle of attack is increased to give a small increase in lift, there will be a large increase in drag. This needs an increase in thrust to match. A stage can be reached where maximum power is required just to match the drag created during landing.

Effect of Sweep Back on Stalling

Swept wing aircraft have the undesirable characteristic of tip stalling. The span-wise movement of the air creates a span-wise boundary layer flow that gradually slows due to friction as it approaches the wing tips. This means that a pool of stagnating air gathers at each wing tip. This thickens the boundary layer at the tips causing it to separate and break away. This stall produces a dramatic fall in lift at the tips. Worse is to come. At higher angles of attack the airflow separates from the leading edge of the wings forming a vortex over the leading edges that moves out towards the wing tips where it separates to join the trailing vortex. This is called the 'Ram's Horn Vortex' and it occurs on swept wing aircraft at higher angles of attack. The Ram's Horn Vortex' can separate from the wing just prior to the tip. If it does this it will affect the ailerons and make lateral control of the aircraft difficult.

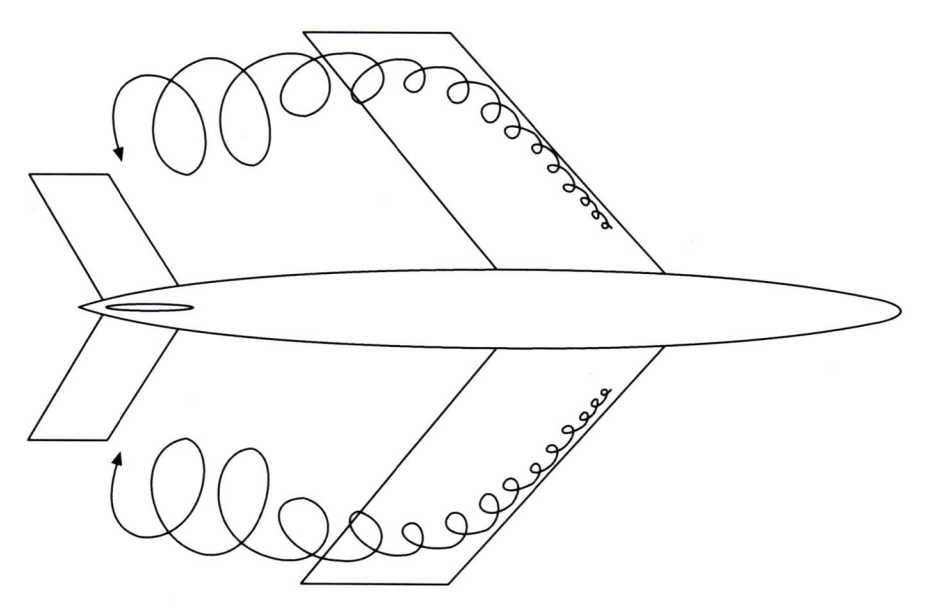

Figure 2.50 - Leading Edge Separation (Rams Horn Vortex)

Experiments were carried out using forward swept wings to alleviate the problems of tip stalling. This configuration saw the span-wise flow moving towards the root instead of the tip and the Ram's Horn Vortex reversed to spiral inwards. This meant that stall would be first encountered at the wing root end leaving the tips clear and aileron control available. The aircraft needed a horizontal stabiliser and pitch control surface forward of the wings because the vortex flow would interrupt the operation of a rear mounted stabiliser. This was provided by using a variable angle Canard forward of the wings.

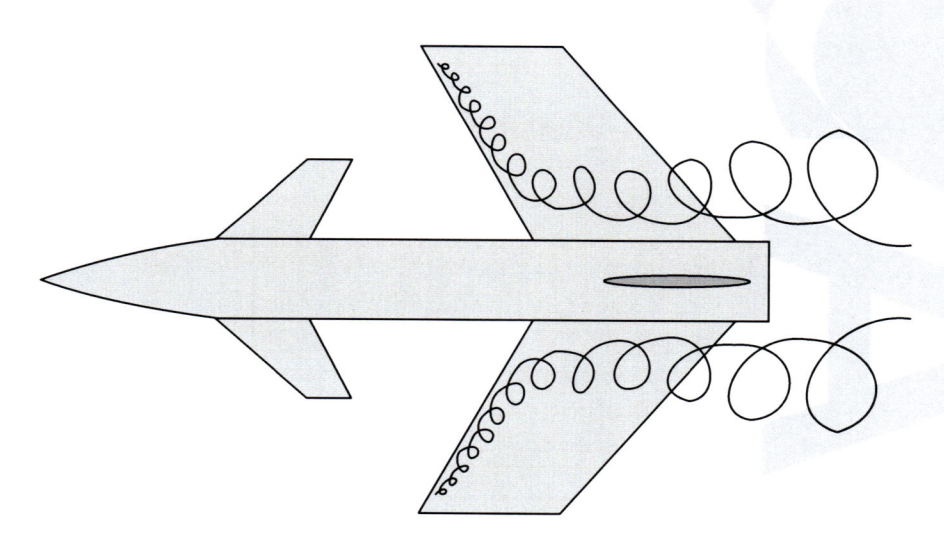

Figure 2.51 - Effect of Forward Sweep on Rams horn Vortex

A number of methods have been employed to alleviate the tip stalling on swept wing aircraft:

1. Boundary layer fences were fitted chord-wise at the outer half of the wing. These were designed to halt the outflow of the boundary layer before it stagnated into the tip area.

2. Leading edge slots were used to duct some of the higher energy air from underneath the wing to mix in with and re-energise the sluggish span-wise boundary layer.

3. Suction was employed to draw the sluggish boundary layer away through ducts in the wing top surface.

4. Blowing has been used to inject pressurised air into the stagnating boundary layer.

5. Vortex generators have been used to re-energise the boundary layer. These are pairs of stubby, aerofoil shaped vanes that create small spinning vortices over the wing designed to draw higher energy air down into the stagnating boundary layer.

6. Leading edges have been designed with notches or 'saw-tooth' profiles to create a tight chord-wise vortex before the stagnating boundary layer can reach the tips. They act like an invisible boundary layer fence.

You can see that a lot of effort has been put into trying to alleviate the tip stalling characteristic of the swept wing. Washout and reducing the tip camber is also commonly employed to delay the airflow separation at the tips.

High Speed Stall

This condition is often referred to as a 'Shock Stall'. It occurs when the V1 airflow velocity reaches Mach 1. When this occurs, a normal shock wave forms on the top of the wing at the point of maximum curvature. Because of the acceleration of the airflow over the camber of the wing, it can reach Mach1 long before the aircraft reaches Mach 1. The air passing through the shock wave, decelerates, increases in pressure and becomes turbulent. This effectively destroys the lift over the wing. Hence the name 'shock stall'. It is the chord-wise airflow that creates this shock wave. By sweeping the wing, the chord-wise airflow velocity (V1) is reduced as I have explained earlier. The effect of this is to delay the formation of the shock wave until a higher aircraft speed is reached. The Mach number that the aircraft reaches when a normal shock wave first forms on any curved surface on the aircraft is called the 'Critical Mach No.' for the aircraft.

The speed of sound in air reduces with a drop in absolute air temperature. Temperature falls up to the Tropopause. As the Mach No. is the ratio of the aircraft's true airspeed to the local speed of sound in air it follows that the aircraft's Mach No. gradually rises as it gains altitude even though its true airspeed remains constant. You have examined the effect of air density on the stall. An aircraft's stalling speed increases with altitude. Here we can have a Catch 22 situation. As the aircraft climbs at constant speed, its Mach No, will progressively get closer to the critical Mach value. At the same time the low speed stalling speed is rising. If the aircraft reaches too high an altitude, it can become trapped between the high and low speed stalls. Increase speed and the aircraft shock stalls, decrease speed and it will low speed stall. Pilots refer to this situation as 'coffin corner'.

The subject of high speed stalling and critical Mach numbers is actually a module 11 topic but I have included a brief description of them, as they are a part of the swept wing stall characteristics.

Pitch-Up

Another problem with swept wings is the centre of pressure movement that occurs during a tip stall. Because of the geometry of the wings, the centre of pressure will move forward during tip stalling. This reduces the lever moment that it can exert around the aircraft's centre of gravity causing the nose of the aircraft to pitch upwards during a tip stall. This aggravates the stall condition and can lead to a complete stall.

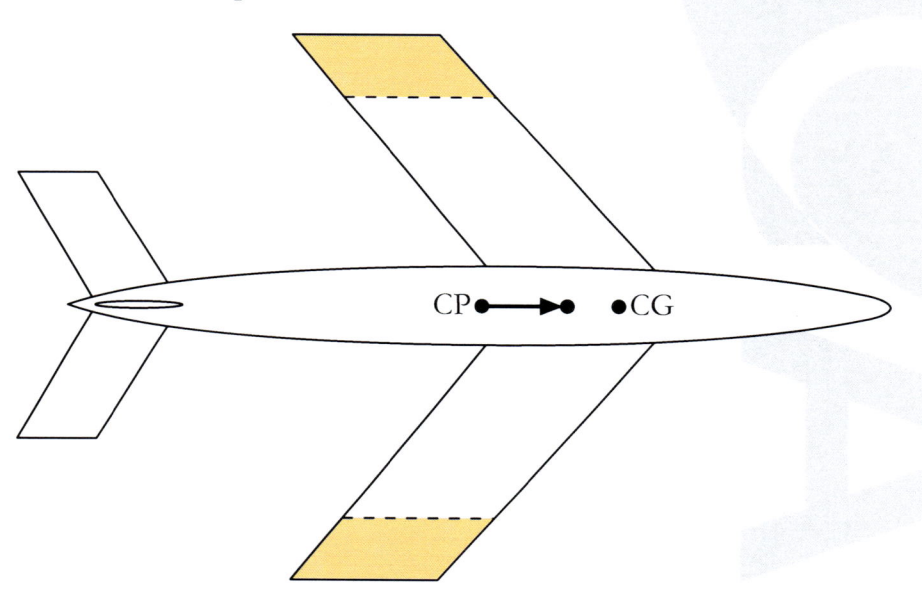

Figure 2.52 - Nose Up Pitch as a Result of Tip Stall

In un-stalled flight, the tip areas have a high coefficient of lift. When the tips stall, the effective lift is transferred inboard. The increase in downwash resulting from this tends to act on the tail-plane pushing in down. This aggravates the nose pitch-up tendency that already exists on swept wing aircraft during tip stall.

Pitch Down

High speed swept wing aircraft experience a nose down tendency when passing from sub-sonic to supersonic flight. This often entails raising the nose using a variable incidence tail-plane. This action creates an additional drag called *trim drag* that adds to the profile drag.

Aerofoil Contamination

Anything that alters the shape or surface finish of an aerofoil will affect its performance. Dirt embedded in grease or oil on the surface will increase skin friction and thus the profile drag. Dents and scratches caused by heavy footed or handed tradesmen will increase the profile drag. Other than these, the type of contamination that comes to mind first is that created by ice, snow or frost.

Ice

Aircraft in flight or on the ground may encounter a number of conditions that can lead to the formation of frozen deposits on the aircraft.

Super-cooled Clouds

Droplets of water can exist in the liquid state at temperatures as low as -40°C. They are unstable and will freeze upon impact with a solid body like an aircraft. The rate at which this type of ice will accrete on leading edges will depend upon the size and quantity of the droplets and the size, shape and speed of the impacting surface. When this type of accretion occurs the ice that builds up on leading edges is opaque, rough surfaced and aerated. This is called *Rime Ice*. One problem that occurs is that water must give up latent heat in order that it may freeze. It does this at the rate of 83 calories per gram resulting in some of the ice melting again and the residual water running back over the aerofoil surfaces and re-freezing at curved surfaces in a form of ice known as *Glaze Ice*.

This type of ice is dense and has high adhesion to surfaces and is difficult to detect on the ground due to its transparency. It can break away in large heavy chunks.

Ice Crystal Clouds

These are clouds that exist at very low temperatures where the moisture in them is frozen into ice crystals. This can impact onto surfaces forming rime and glaze ice deposits.

Freezing Rain

This is another form of super-cooled moisture and will accrete as rime and glaze ice.

Rain and Drizzle

Aircraft may be descending from altitude at a temperature below freezing. If they encounter rain it will freeze and accrete as **Rain Ice** on the surfaces. This ice can be dense and glazed with high adhesive power.

Frost

If the air is humid and the aircraft temperature is below freezing, the moisture will precipitate out of the air onto the aircraft surfaces as **Hoar Frost**. This has high adhesive powers and is rough surfaced.

Snow

Snow may accrete on aircraft either when it falls or when it is blown onto the aircraft by taxiing aircraft or high winds. Snow can also be impacted onto parts of the structure during ground movement of the aircraft.

Affects of Ice Accretion

The components of an aircraft that are affected by ice accretions are:

- Aerofoil surfaces designed to produce lift
- Stabilising devices
- Control surfaces
- Engine inlets
- Propellers
- Rotor blades
- Windscreens
- Air inlets
- Antennae
- Landing gear
- Fuel system vents

We are only concerned with aerofoil contamination in this syllabus so I will limit the description of the effects on aircraft performance to those stemming from aerofoils and components that may have an indirect affect on them. The effects of ice accretions on these components and the degradation in aircraft performance and flight characteristics that result from this are:

1. The surface roughness created by frost and ice deposits on flying surfaces may alter the stalling speed and stalling behaviour of an aircraft.

2. The surface roughness created by frost and ice deposits on the leading edges and flight surfaces will produce additional drag and may reduce lift. The extra weight will require extra lift to support it and this will increase the induced drag. Surface roughness on aerofoils will in particular disrupt airflows and reduce lift.

3. The change in stall speed will change the parameters for manoeuvres such as turning, approach and landing speeds. Lift off speed could also be affected.

4. Accretions of ice may affect the weight and the balance of the aircraft. This will again affect the lift required to maintain safe flight and manoeuvres.

5. Ice accretions can alter the shape of aerofoils reducing their ability to operate within the flight envelope of lift and stall characteristics.

6. The stall angle of attack may be reduced leading to unpredictable stalls before the on board stall warning systems operate.

7. Ice breaking off the fuselage and wings can cause impact damage to the wing and tail leading edges.

8. Ice forming on engine air inlets and propellers can reduce engine efficiency and in latter case can cause vibration. This can reduce available thrust.

9. Ice accretions on air speed and altitude sensing probes will create false readings and may affect automatic flight control systems.

10. The aircraft control surfaces may become seized or unbalanced.

11. The disruption to air flow caused by ice accretions may alter the centre of pressure movement and its position.

The contamination of an aerofoil with ice will increase drag, reduce lift, change stall characteristics and reduce the stall angle of attack. The most significant risk comes from accretions of rime ice on the leading edges where most ice will accrete in flight. Glaze ice may however freeze back on the camber forming ridges that will cause separation of flow.

The rough surface of hoar frost will significantly increase skin friction.

This marks the end of chapter two and I suggest that you take a break before trying out your powers of recall in the test that follows. See if you can score over 75%. Good luck!

Revision

Aerodynamics

Questions

1. When the airflow velocity starts to decrease over the top surface of a cambered aerofoil the:

    a. pressure decreases and the lift increases

    b. pressure increases and the lift increases

    c. pressure increases and the lift decreases

2. The layer of air adjacent to the surface of an aerofoil that contains flow velocities that are lower than that of the free-stream airflow is called the:

    a. boundary layer

    b. downwash layer

    c. adverse gradient

3. The transition point is where the:

    a. airflow divides at the leading edge

    b. flow separates at stall

    c. laminar flow becomes turbulent

4. Induced drag is:

    a. proportional to the square of the aircraft speed

    b. inversely proportional to the square of the aircraft speed

    c. directly proportional to the aircraft speed

5. When the boundary layer flow changes from laminar to turbulent its depth:

    a. increases

    b. decreases

    c. remains unaltered

6. An aspect ratio of 5 means the:

    a. mean chord is five times the span

    b. span is five times the mean chord

    c. span is the square root of the area

7. The lift created by a wing is said to act through the:

    a. centre of gravity

    b. centre of pressure

    c. transition point

8. The downwash off a wing will:

    a. increase the angle of attack

    b. not affect the angle of attack

    c. decrease the angle of attack

9. A stall on a straight rectangular wing commences:

    a. at the root first

    b. at the tip

    c. equally along the span

10. When a wing is in the un-stalled condition the Centre of Pressure (CP) will move:

   a. rearwards as the angle of attack increases

   b. forwards as the angle of attack increases

   c. forwards as the aircraft speed increases

11. At the onset of a stall the Centre of Pressure (CP) will:

   a. move forwards

   b. not be affected

   c. move rearwards

12. As aircraft speed increases the:

   a. transition point moves forward

   b. centre of pressure moves forward

   c. separation point moves rearwards

13. Profile drag is dependent upon:

   a. shape

   b. lift

   c. downwash

14. Compared to a straight wing the stalling angle of attack of a delta wing is:

   a. lower

   b. similar

   c. greater

15. As the altitude of an aircraft increases its stalling speed will:

   a. decrease

   b. increase

   c. remain the same

16. The air velocity over a swept wing that contributes to lift is:

   a. higher than the aircraft speed

   b. lower than the aircraft speed

   c. the same as the aircraft speed

17. The trailing vortex on a wing having a taper ratio of zero will be:

   a. near the wing root

   b. at the wing tip

   c. spread evenly along the span

18. The onset of a stall on a rearward swept wing aircraft will occur:

   a. at the wing roots

   b. equally along the span

   c. at the wing tips

19. At low airspeeds the coefficient of lift of a slender delta wing:

   a. decreases

   b. increases

   c. remains unaltered

20. The maximum coefficient of lift of a swept wing compared to a straight wing is:

   a. higher

   b. similar

   c. lower

21. The angle of attack of a wing is the angle between the chord line and the:

   a. relative airflow

   b. longitudinal axis of the fuselage

   c. mean camber line

22. Washout is a term that refers to an:

   a. increasing angle of incidence towards the wing tip

   b. decreasing angle of attack towards the wing tip

   c. decreasing angle of incidence towards the wing tip

23. The angle of incidence will:

   a. decrease with an increase in downwash

   b. increase when the aircraft nose lifts

   c. never change

24. In level flight the aircraft's weight is the product of:

   a. mass and velocity

   b. mass and the acceleration due to gravity

   c. mass and the lift force

25. As the weight of an aircraft decreases in flight its stalling speed will:

    a. decrease

    b. increase

    c. stay the same

26. The wing tip vortices contribute to:

    a. induced drag

    b. profile drag

    c. trim drag

27. As the airspeed increases the total drag on an aircraft will:

    a. initially increase and then decrease

    b. increase

    c. initially decrease and then increase

28. As the speed of an aircraft increases the profile drag will:

    a. initially decrease and the increase

    b. increase

    c. initially increase and then decrease

29. The line of action of the weight of an aircraft always acts:

    a. parallel to the normal axis of the aircraft

    b. vertically downwards through the centre of gravity

    c. vertically downwards through the centre of pressure

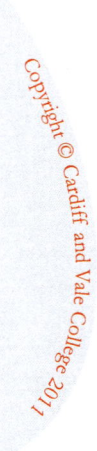

30. When a swept wing aircraft experiences the onset of stall it will:

    a. pitch nose up

    b. pitch nose down

    c. remain straight and level

31. If the weight of an aircraft is increased its minimum drag speed will:

    a. decrease

    b. not alter

    c. increase

32. The line of action of the lift force always acts at right angles to the:

    a. chord line

    b. relative airflow

    c. camber line

33. Interference drag is created by:

    a. frontal area

    b. component joints

    c. surface finish

34. The total drag acting on an aircraft is the sum of:

    a. induced drag and profile drag

    b. induced drag and form drag

    c. profile drag and trim drag

35. **A high aspect ratio wing will create:**

    a. low induced drag

    b. high induced drag

    c. increased downwash

36. **If the air density decreases the:**

    a. lift increases

    b. drag does not alter

    c. drag decreases

37. **Washout is provided to:**

    a. reduce interference drag

    b. prevent span-wise flow

    c. reduce induced drag

38. **The maximum lift/drag ratio on a straight wing will occur at an angle of attack of:**

    a. 16°

    b. 4°

    c. 8°

39. **The minimum total drag in flight occurs when:**

    a. profile drag equals induced drag

    b. induced drag reaches a minimum

    c. the coefficient of lift is at its maximum

40. Ice forming on the wings will:

   a. reduce induced drag

   b. increase induced drag

   c. increase the lift coefficient

CHAPTER TWO
AERODYNAMICS

Revision

Aerodynamics

Answers

1. **C**
2. **A**
3. **C**
4. **B**
5. **A**
6. **B**
7. **B**
8. **C**
9. **A**
10. **B**
11. **C**
12. **A**
13. **A**
14. **C**
15. **B**
16. **B**
17. **A**
18. **C**
19. **B**
20. **C**
21. **A**
22. **C**
23. **C**
24. **B**
25. **A**
26. **A**
27. **C**
28. **B**
29. **B**
30. **A**
31. **C**
32. **B**
33. **B**
34. **A**
35. **A**
36. **C**
37. **C**
38. **B**
39. **A**
40. **B**

Theory of Flight

Relationship Between Lift, Weight, Thrust & Drag

Lift

When an aircraft is parked on the ground its weight acts through the aircraft wheels onto the ground where it is balanced by an equal and opposite reaction force in accordance with Newton's third law. To keep an aircraft in the air in level flight at constant altitude, however, it is necessary to generate a vertical, upward force to exactly counteract the weight of the aircraft. We call this the **lift** force.

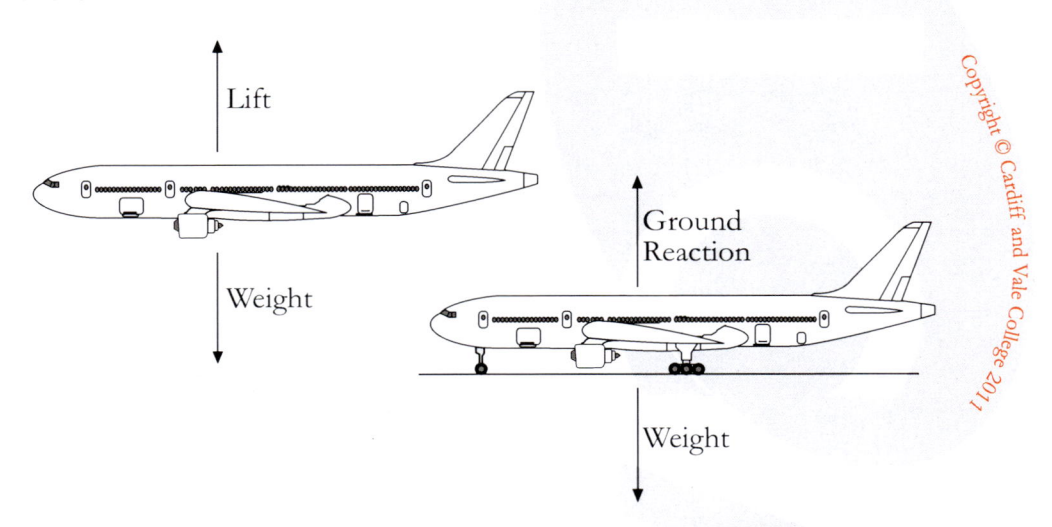

Figure 3.1 - Supporting Aircraft Weight on the Ground and in the Air

The lift an aircraft needs to rise off the runway depends upon its all up weight. The heavier the aircraft is the more the wings have to lift and the higher the take-off speed has to be. Most large airports have runways two or three miles long to allow aircraft to reach the required speeds. The altitude and ambient conditions also affect the take-off speed. Low air density indicates more speed to develop the necessary lift to overcome the weight. The speed at which an aircraft is committed to lift off is called V1. Prior to this the take-off can be aborted. Soon after V1, the aircraft nose is rotated up to present the wings at a suitable angle of attack to generate the lift-off force. The speed at which this occurs is called V_R.

There are occasions when an aircraft is not flown straight and level. Climbing, diving and turning are examples of this. During these manoeuvres the lift force will not be acting vertically upward and may not exactly balance the aircraft weight. The lift force will also vary with aircraft speed, altitude and ambient conditions. We need to find a definition to describe the lift force.

Lift is a force that acts at right angles to the line of flight or relative airflow (RAF) through the *centre of pressure (CP)* of the aircraft wings.

Weight

The weight of a body is a force that always acts vertically downwards through its centre of gravity. Regardless of the position or attitude of the body its weight will continue to act vertically down. The weight of an aircraft will vary in flight as fuel is consumed but not, as some people will have it, when the flight rations are consumed! We should find a clear definition to describe the force we call weight in relation to an aircraft.

Weight is a force that acts vertically downwards through an aircraft's *centre of gravity* regardless of the attitude the aircraft is in.

Thrust

If we wish to fly an aircraft at a constant speed we have to produce a forward acting force to overcome and then balance the pull of the rearward drag force created by the air resisting the passage of the aircraft through it. We call this the *thrust* force. This force is normally provided from reaction thrust created by the engines. There will be occasions when there is a need to accelerate, decelerate, climb, dive or turn an aircraft and the thrust force will no longer balance the drag force. In a climb, a component of the thrust force is required to complement the lift force to support the weight. Again, we need to find a definition to describe the thrust force.

Thrust is a forward acting force that acts parallel to the line of flight and is arranged symmetrically about the longitudinal axis of the aircraft.

Drag

The drag force acts to resist the passage of the aircraft through the air. The force varies with aircraft speed, altitude, the ambient conditions and the value of the lift force. There will be occasions when the drag force does not equal the force of thrust and the aircraft will accelerate or decelerate accordingly. We need a definition to describe the force.

Drag is the force that opposes the forward motion of an aircraft and is considered to be a rearward force that acts along a line that is parallel to the line of flight.

Arrangement of the Four Flight Forces

If an aircraft is to fly straight and level at a constant forward speed and altitude then the lift must equal the aircraft weight and the thrust must equal the drag. If these forces are in balance then the aircraft will be in a state of equilibrium.

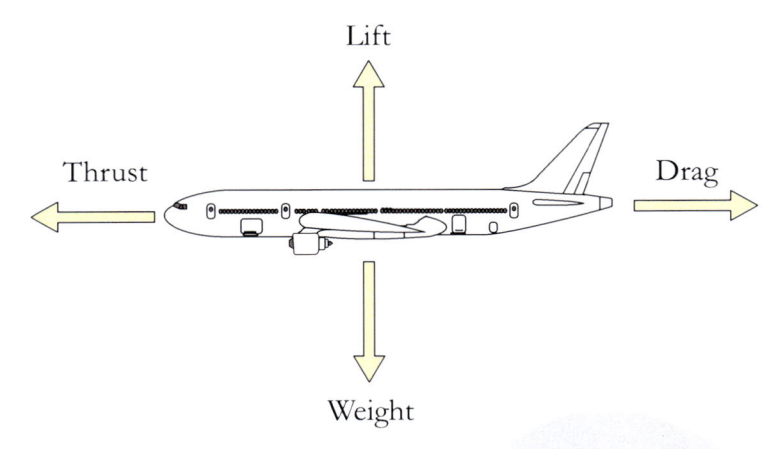

Figure 3.2 - Forces Acting on an Aircraft in Level Fight

The angle of attack of the wings is adjusted to provide the lift to balance the weight at the selected altitude and the engine thrust is adjusted to balance the drag at the selected airspeed. If the airspeed is low then a higher angle of attack is required to maintain the lift and thrust will be adjusted to balance the drag and maintain the chosen airspeed. If the airspeed is increased then the angle of attack must be reduced to prevent the lift increasing above the value required to maintain the altitude. The thrust will be increased to maintain the higher airspeed and to balance the increase in drag at that airspeed.

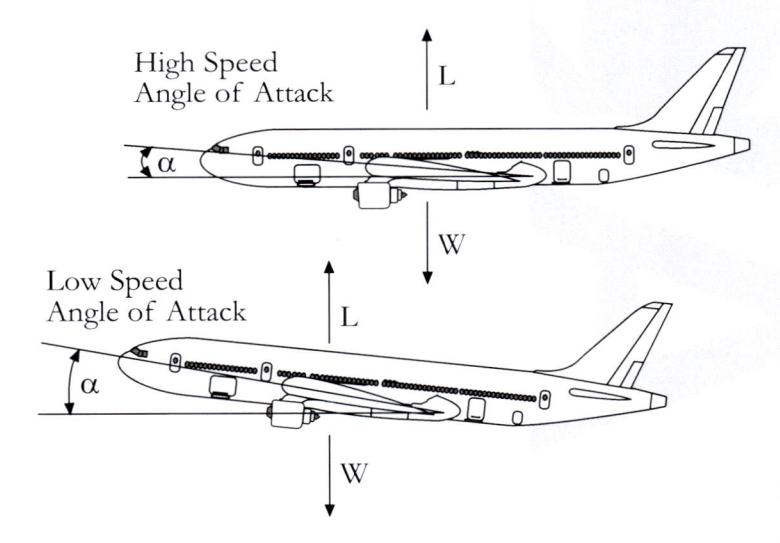

Figure 3.3 - Maintaining level Flight

If during level flight the lift force were to be reduced so that it did not equal the weight, the aircraft would sink and lose altitude. Conversely, should the lift force be increased to exceed the weight, the aircraft would rise and gain altitude.

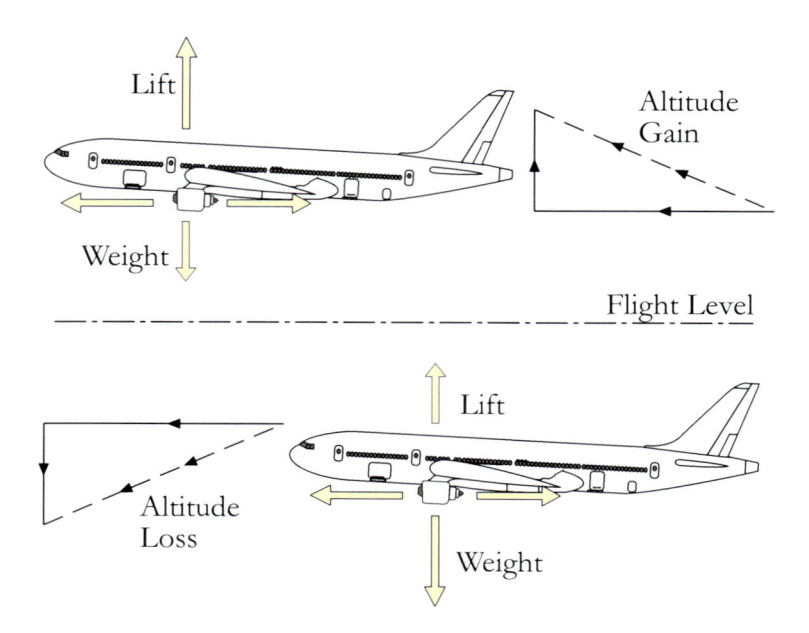

Figure 3.4 - Effects of Changes in Lift and Weight

The weight of an aircraft changes in flight because it consumes fuel so the lift has to be adjusted. Reducing the angle of attack will **reduce the lift until it balances the weight again**. This would require a reduction in thrust to balance the reduced drag.

If the thrust were to be increased above the value of the **drag force, the aircraft would accelerate. Lift would increase so the angle of attack would need to be** reduced if altitude is to be maintained. The aircraft would **then accelerate until** the increasing drag force matched the new thrust setting. Conversely, if thrust were to be reduced below the value of the drag force, the aircraft would decelerate and lift would again have to be adjusted to balance the **weight**. The aircraft would then decelerate until the reducing drag force **matched the new** thrust setting.

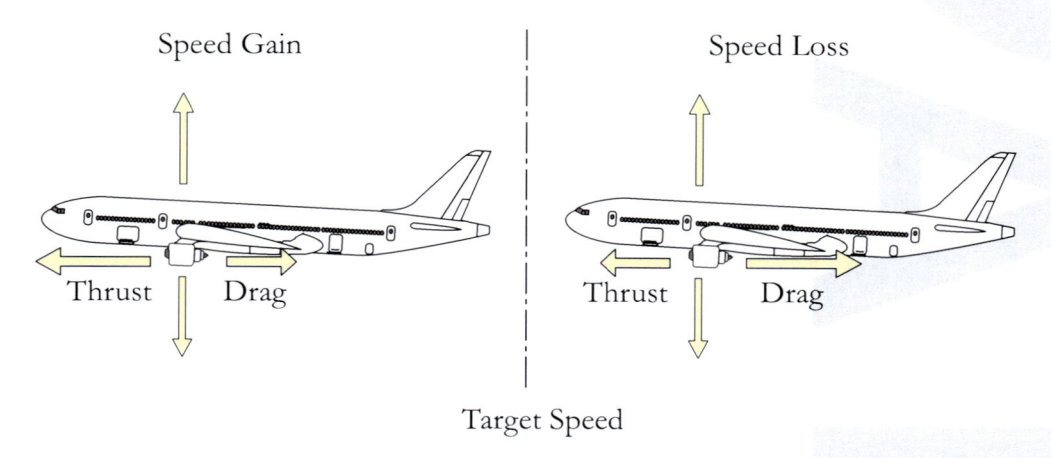

Figure 3.5 - Effects of Changes in Thrust and Drag

Relationship Between the Four Forces

We have already examined the relationship between lift and lift-induced drag in chapter 2. Any change in the lift will produce a proportional change in induced drag. We also know that there is an optimum value for the lift/drag ratio in relation to the angle of attack.

The lift acts through the centre of pressure and we know that this point moves in relation to the angle of attack. This means that the line of action of lift also moves. The weight acts through the centre of gravity and this point can also move depending on the changes in aircraft weight and loading. As the opposing lines of action of lift and weight do not correspond they form a 'couple' and there will be a resultant turning or pitching moment. The centre of gravity is normally positioned ahead of the centre of pressure so an increase in lift will strengthen the couple and exert a nose down turning moment and a decrease in lift will weaken the couple and produce a nose up movement.

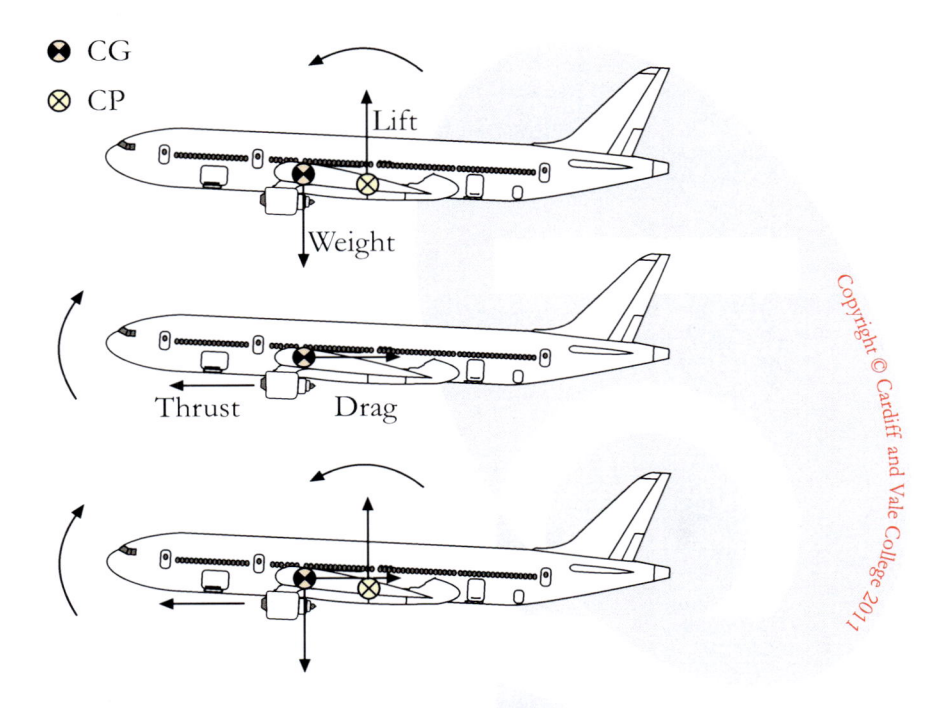

Figure 3.6 - Arrangement of Flight Forces

The line of action of thrust is normally positioned below the line of action of drag forming a 'couple' that will cause a nose up tendency. Changes in either of the two forces will create a turning moment. If the line of action of thrust is below the line of action of drag an increase in thrust will strengthen the couple and create a turning moment, pushing the nose up.

To achieve a state of equilibrium, the two couples must balance. If thrust is increased, the nose will pitch up. The increase in airspeed will create more lift and that will pitch the nose down. Drag increases but this will not create a pitching moment providing its line of action passes through the centre of gravity.

A balance will be created. If the engines were to fail, the aircraft nose would pitch down under the influence of the lift weight couple. If the design is right the nose will move down to adopt an optimum angle for the aircraft to glide. If the pitching moment were different, resulting in a nose up movement following an engine failure, the aircraft could then slow into a stalled condition. Not nice!

This is intended as a brief description only of the relationships between the lines of action of the forces. As these have a direct influence on the stability of an aircraft I will be entering into more detail later in this chapter and in Chapter 4.

No Power Gliding

We have examined an aircraft in level flight with all the flight forces in equilibrium. If the engine/s were to fail we will have a situation where one of the flight forces, namely thrust, will have been removed. We would now require that the lift, drag and weight forces reach equilibrium. If the thrust is removed when the aircraft is flying straight and level the drag is no longer balanced by it and the aircraft will slow down. To maintain flying speed the aircraft nose will have to be lowered. The drag will now be opposed by a component of the weight. The weight and the drag components act parallel to the direction of flight or glide path and oppose each other. If the drag is increased for any reason, by lowering the flaps or undercarriage for example, a greater component of weight will be required to balance it so the glide path will have to be made steeper.

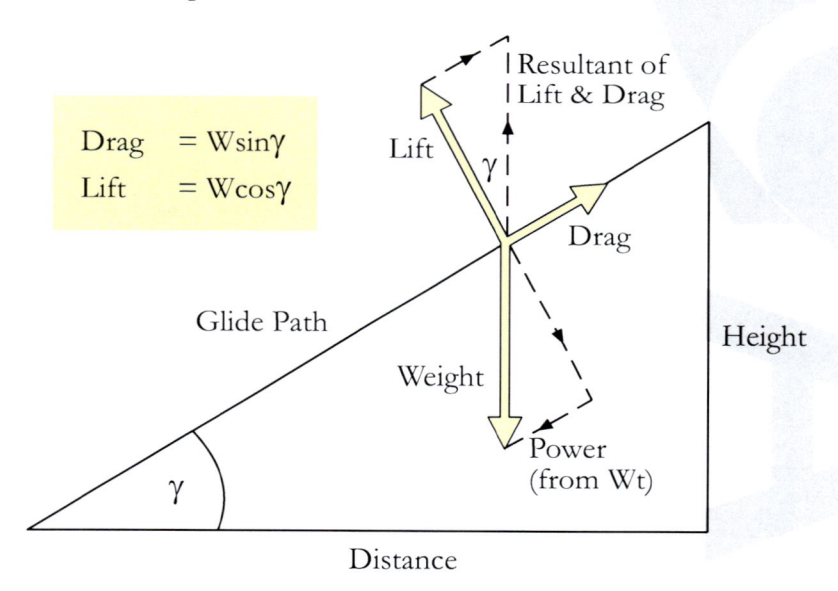

$$\text{Drag} = W\sin\gamma$$
$$\text{Lift} = W\cos\gamma$$

Figure 3.7 - Forces in a Glide

If everything is right, the aircraft weight will be balanced by the resultant of the lift and the drag forces. The lift force acting perpendicular to the glide path will be tilted forwards and the drag vector will be parallel to the line of flight. We no longer have an engine to overcome this drag so we must utilise the potential energy of the aircraft height. The weight still acts vertically down and this produces a component of weight that acts along the glide path. We use this as power to balance the drag. Instead of an engine we are using the aircraft weight for propulsion. A glider does this.

At this point a decision has to be made. We can opt for staying in the air as long as possible or, we can opt to gain the maximum distance. Simply we must decide between endurance and range.

If we choose endurance we wish to remain in the air for as long as possible so we want the minimum rate of descent or sink. To achieve this we need to use the aircraft potential energy sparingly and this requires that we adopt a lower than normal glide angle to reduce the power we get from the weight. We will have increased lift because the lift vector will not be tilted so far forward. The aircraft speed required will be around three-quarters of the value normally identified for a maximum range glide. This will minimise the 'sink' rate. We do need to overcome drag so we adopt a forward speed that minimises the power from weight vector whilst balancing drag. We are keeping the 'sink' rate low but this does not mean we will gain any bonus in forward distance. Think of it as two people standing on a high diving board. One jumps off and deploys a parachute while the other dons a pair of wings and glides. The former will develop a lot of lift and float slowly down to the pool not travelling very far forward. The latter will maintain the highest possible lift/drag ratio and will reach the far end of the pool long before our parachutist hits the water.

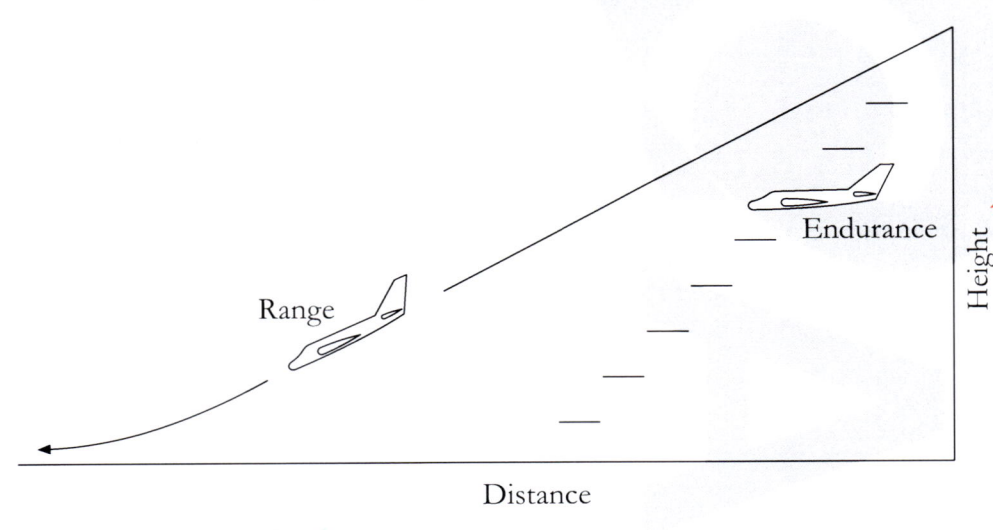

Figure 3.8 - Gliding for Range or Endurance

If we choose range we wish to cover as much forward distance as possible. To do this we must choose an angle of glide that will establish an angle of attack that gives the highest possible lift/drag ratio. We are flying the maximum distance by flying with minimum drag.

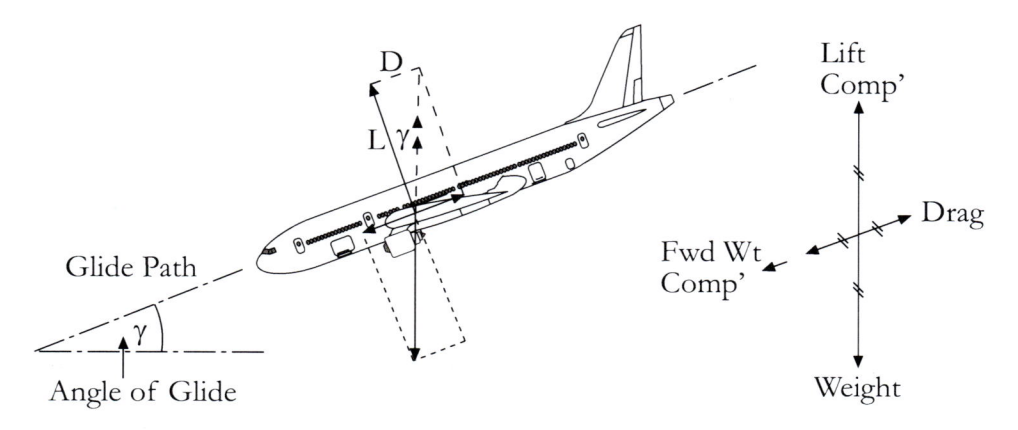

Figure 3.9 - Angle of Glide

The angle of glide is represented by the Greek letter gamma (γ). This is the same as the angle formed between the resultant of lift and drag and the lift vector. Looking at the illustration you may see that the **tangent of γ is** Drag/Lift (Opposite/Adjacent).

$$\text{Tan } \gamma = \frac{\text{Drag}}{\text{Lift}}$$

To get a low value means increasing the lift and reducing **the drag**. Simply put this means a high lift/drag ratio or C_L/C_D. So, to get maximum **range** in a glide we need an angle of glide that produces the highest possible **lift/drag ratio**. A typical lift/drag ratio for an aircraft in a glide is about 10:1. Let us see what glide path angle suits this.

$$\text{Tan } \gamma = \frac{\text{Drag}}{\text{Lift}} = \frac{1}{10} = 0.1 = 5.7°$$

If we attempt to glide at an angle less than the optimum angle by pulling the aircraft nose up in an attempt to extend the range then the descent will become steeper because the drag will increase, the speed will reduce and the range will reduce. Changing the angle of attack of the wings in a glide will alter the lift/drag ratio and the airspeed. If you travel at a speed over or under the correct speed it will result a steeper descent and reduce the range. Too fast and the range reduces, too slow and the range reduces even more. The steepness of a glide is dependent on the lift/drag ratio.

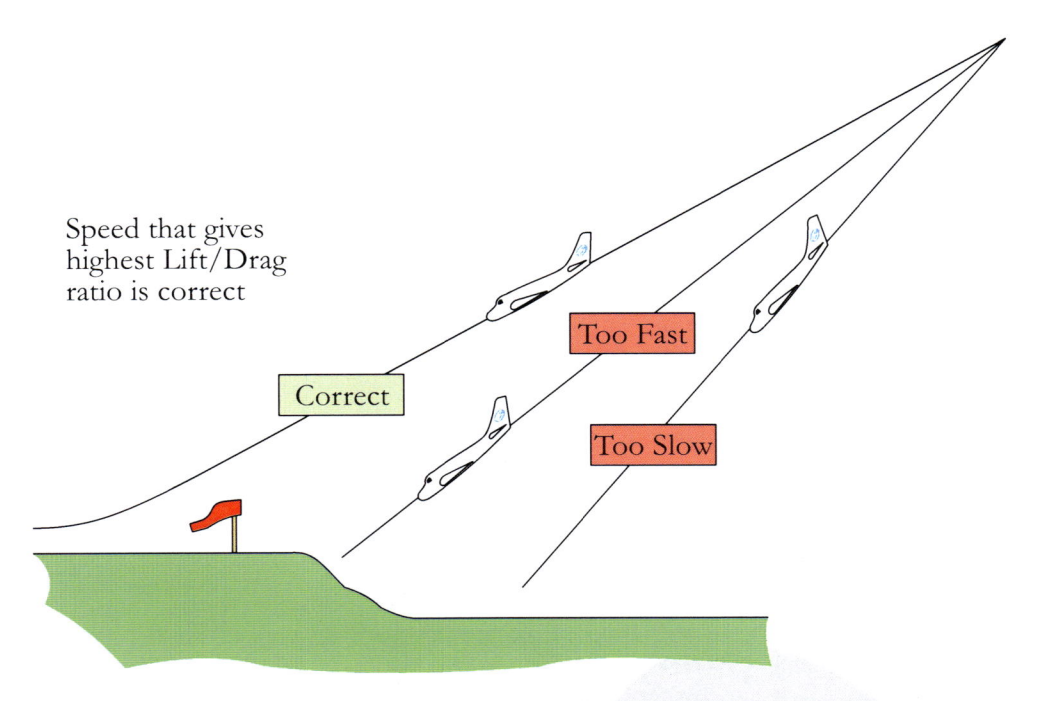

Figure 3.10 - Speed for Best Lift/Drag Ratio

Effect of Wind

The effect of a wind will not necessarily affect the endurance of a glide but it will affect range. A headwind will make the descent steeper whereas a tailwind will reduce it. At the gliding airspeed for range, a headwind will reduce the distance the aircraft covers in relation to the ground. However, if the aircraft speed is increased in a headwind although the rate of descent will increase the aircraft will be able to penetrate into the headwind and cover more distance. A tailwind will extend the distance covered. In fact, if the aircraft speed is reduced slightly in a tailwind it will reduce the rate of descent and allow the tail wind to blow the aircraft further forward. In no wind conditions maximum range is always achieved by flying at the highest lift/drag ratio.

A question often comes up about the effect of headwinds on an aircraft. The true airspeed of an aircraft is the speed of the aircraft in relation to the air. This does not indicate its actual speed in relation to the ground. ***Airspeed*** and ***groundspeed*** are rarely the same. So, an aircraft flying into a headwind will cover less distance across the ground than it would in still air. The sea level airspeed will be higher than the groundspeed. Conversely, it will travel further across the ground in a tailwind. The ground speed will be higher than the sea level airspeed.

Figure 3.11 - Effect of Wind on Glide Range

Effect of Weight

The weight of an aircraft does not affect the gliding angle. The lift must be increased but this has the effect of changing the other force vectors so that the overall geometry of the forces remains unchanged and thus the glide angle is unaltered. The gliding speed is affected however, a reduction in weight means that the ideal airspeed will be lower. In this respect the gliding range will be unaffected. Weight does affect endurance however. The rate of descent will increase so endurance will decrease with an increase in weight.

Glide Ratio

When gliding for range the ratio of distance covered to height lost is known as the glide ratio. It is similar to the lift/drag ratio. If the lift/drag ratio were typically 10:1 then we can say that for every one thousand feet of descent we will travel forward a distance of ten thousand feet. If we were at an altitude of six thousand feet, for example, at a glide ratio of 10:1 our range to landing will be sixty thousand feet or twenty thousand yards which is, just over eleven miles.

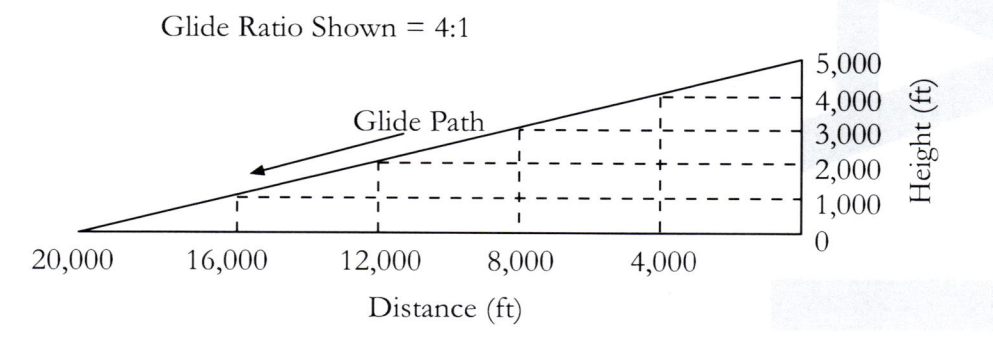

Figure 3.12 - Glide Ratio

Steady State Flights - Performance

Steady state flight includes level flight, climbing, powered descent and turning at constant speed. We could have included power off gliding but we have already dealt with that.

Straight and Level Flight

When an aircraft is in steady level un-accelerated flight the aircraft is trimmed so that lift equals weight and thrust equals drag. We have examined the arrangement of these four forces and know that the lines of action of each differ.

Looking again at lift and weight. Let us remind ourselves that the lift acts through the centre of pressure, the weight acts through the centre of gravity. This forms a couple which is defined as two opposing forces that do not have the same lines of action. The effect of these forces is to produce a turning moment. The turning moment will result in a nose up or a nose down tendency dependent on whether the centre of gravity is in front of, or behind the centre of pressure. Normally it is in front. In this case an increase in lift will strengthen the lift weight couple and produce a nose down pitching moment and vice-versa. The centre of pressure position moves forward as the angle of attack is increased. This will tend to reduce the turning moment but the increase in lift will restore it. As the angle of attack is reduced the centre of pressure moves backwards this will tend to increase the turning moment but the decrease in lift will reduce it.

The thrust and drag force lines of action also produce a couple. Normally an increase in thrust will pitch the nose up. If the throttle is pulled back, the thrust drag couple weakens and the nose will drop.

When the aircraft is in steady level flight the pitching moments of the two couples will balance each other so there should be no residual turning moments.

The tail-plane or, horizontal stabiliser is fitted to counter any residual pitching moments should they arise. Because it is fitted some distance aft of the centre of gravity the tail-plane exerts a powerful restoring moment. If the tail-plane has to exert a downward restoring force this will add to the aircraft's apparent weight. Fortunately the forces experienced by the tail-plane are relatively small due to the long moment arm.

We have already examined the factors to be considered when varying the airspeed in level flight. Recall that a reduction in airspeed necessitates an increase in angle of attack to restore lift. At cruising speed, the angle of attack is reduced to attain the balance between lift and drag. Normally cruising speed will be set around the minimum drag speed V_{MD}. The angle of attack will be around three to four degrees giving the highest lift/drag ratio. The speed changes are made possible by increasing or decreasing the thrust to achieve the required acceleration or deceleration and to match the drag at the selected airspeed.

As the aircraft weight reduces in flight, the lift has to be adjusted to match it. This necessitates either a reduction in angle of attack or aircraft speed. The minimum drag speed V_{MD} will also reduce with a reduction in aircraft weight.

At altitude the air density reduction will affect the airspeed, lift and drag so the relationship between the angle of attack and the lift does not alter. However, the aircraft weight does not alter with altitude so the angle of attack or the airspeed may have to be increased to maintain the balance of lift and weight. Lift is reduced at altitude due to the reduced air density.

Climbing

During a climb the aircraft will be gaining potential energy due to height so a higher thrust will be required than in level flight. To climb at a steady speed the thrust has not only to overcome the drag but it also has to lift the weight of the aircraft to achieve the vertical speed and height. This is referred to as the rate of climb. In a steady climb, lift is less than the aircraft weight. The rate of climb will be determined by the power applied and the angle of the climb.

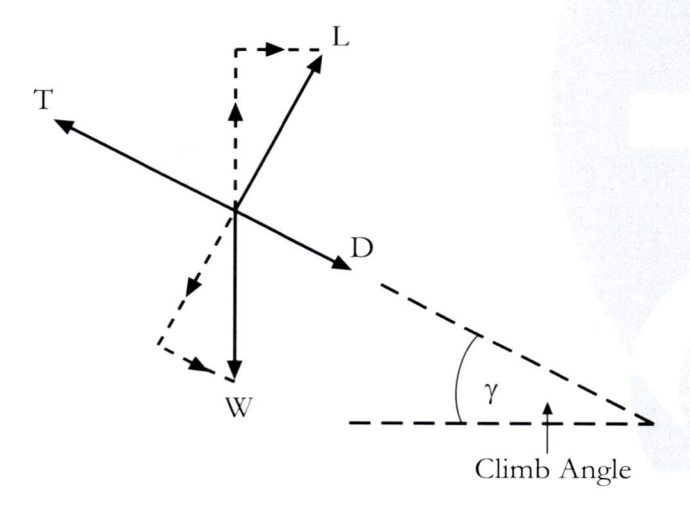

Figure 3.13 - Climb Angle

The angle of climb γ is derived by the following formula:

$$\sin \gamma = \frac{\text{Thrust - Drag}}{\text{Weight}}$$

The climbing speed is dependent on the power available to create thrust. The power should be the maximum possible to overcome the drag and leave a surplus to climb the aircraft. If you examine the illustration you may see that the lift vector is tilted rearwards and that the vertical component of lift L_{vert} does not balance the weight of the aircraft. If you look again, this time at the thrust, you may see that there is a vertical component of thrust T_{vert} acting with the vertical component of lift L_{vert} to overcome the weight. Imagine that if the aircraft were in a vertical climb, there would be no vertical lift vector and the climb would depend entirely on the vertical thrust force.

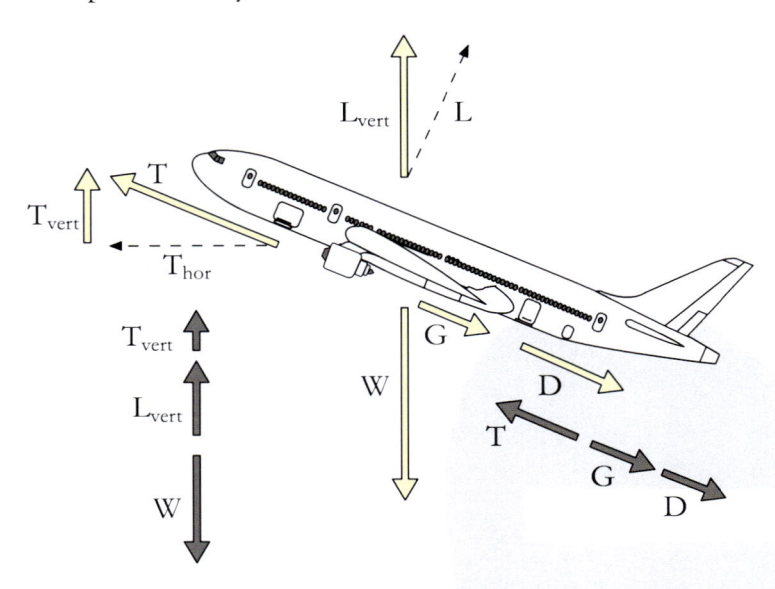

Figure 3.14 - Forces on an Aircraft in a Steady Climb

Look again at the illustration and you may see that the weight has a rearward acting component that I have identified as G. This acts with the drag force to resist the forward movement of the aircraft. The resultant thrust force T has to overcome this. So, to sum up, the engine power required has to produce enough thrust to overcome the drag and the rearward acting component of weight and make up the deficit between the vertical lift component and the weight and then leave enough excess power to climb the aircraft. Do remember one fact that examiners like to test you on. Lift is always less than weight in a steady climb.

As altitude increases the power available to climb the aircraft may be reduced due to falling air density. The drag on the aircraft will also reduce but the weight will not, nor will the energy required to increase the potential energy by height. There will be a ceiling where the maximum available engine power will only just match the requirements. Above this a climb cannot be continued.

Powered Descent

If thrust is available during a descent it will balance some of the drag. This means that the weight component required is less than that required in a glide to maintain airspeed.

The angle of descent will be less than a no power glide and the rate of descent will be reduced. In a steady, controlled descent the objective is to maintain a steady forward speed by balancing the sum of the thrust and the forward acting components of lift and weight to the drag force. To maintain a constant speed, the thrust has to be reduced or the angle of descent reduced to achieve the correct balance between these forces. The steeper the angle of descent the greater will be the forward component of weight so thrust will have to be further reduced. In a way it is similar to setting up the glide path angle for maximum range. A high lift/drag ratio would be the aim. If a rapid descent is required and range is not important then increasing the angle of descent will allow the forward component of weight to increase and speed will increase until the increasing drag balances it again.

Commercial aircraft often employ the cruise descent in that they slightly reduce the thrust and lower the aircraft nose to maintain airspeed while they are several miles out from the airport. This permits a gradual descent to be made whilst still maintaining the cruise airspeed. It is a time saving manoeuvre.

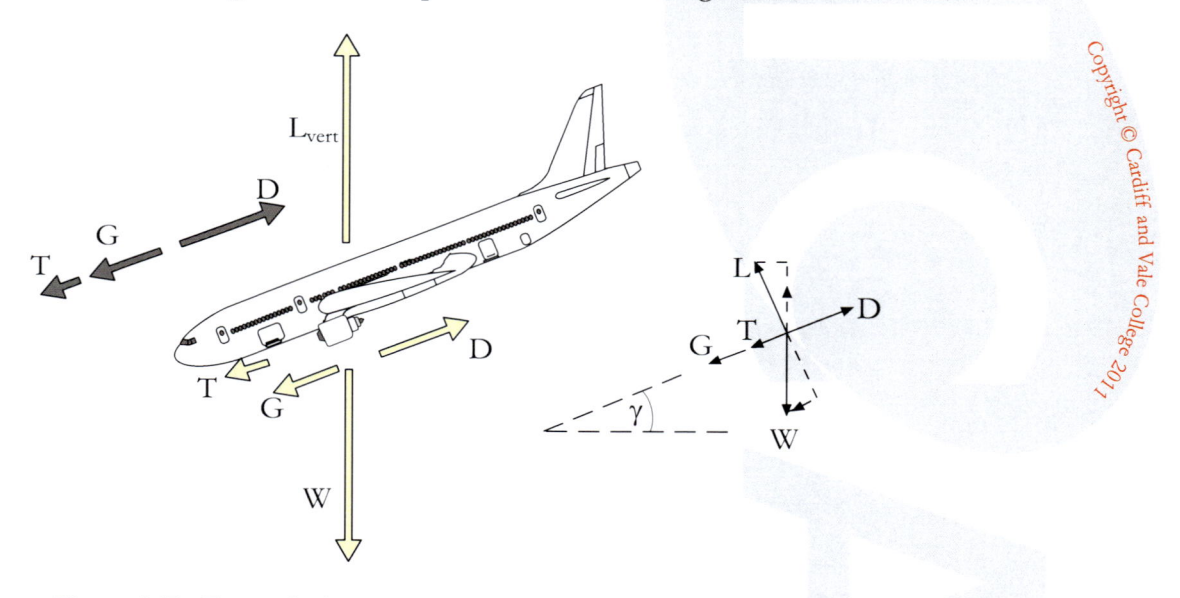

Figure 3.15 - Forces Acting on an Aircraft in a Steady Powered Descent

If the engine power is increased, the aircraft will begin to flatten out. The thrust will balance out more of the drag so the component of weight acting along the glide path need not be so great. As power is increased the pitch attitude of the aircraft will move to nose up and the rate of descent will decrease. With sufficient increase in power the aircraft will pull out of the descent.

The Sideslip

I will include this manoeuvre though it is actually an out of balance condition. When an aircraft is banked so that one wing is higher than the other the lift vector remains perpendicular to the lateral axis of the aircraft. This vector is no longer perpendicular to the ground. The weight vector however, continues to act vertically down. The resultant of the lift and weight vectors is a force that causes the aircraft to be pushed sideways, or sideslip in the direction of the lower wing. Presenting the side of the aircraft to the airflow increases the drag, reducing the lift/drag ratio, so the aircraft will descend. If the aircraft is already descending this manoeuvre will make the descent steeper. The force imposed by the airflow on the side of the aircraft fin and fuselage area aft of the centre of gravity will cause the aircraft to yaw and, unless corrected with opposite rudder, this will cause the nose to drop into a steep angle of descent.

Theory of the Turn

Centripetal & Centrifugal Forces

In order that you may understand the principle of turning we must examine some of the physics governing rotating bodies. This topic is comprehensively covered in module two but in case you have not studied this I will give you a brief introduction to it, sufficient for our purpose here.

If you have ever swung an object around on the end of a cord, you will know that the cord appears to be trying to pull itself out of your hand. If you consider what is happening, you could equally say that while the object was in motion in a circular path, you have to exert a constant pull on the cord to keep it in your hand.

From Newton's first law of motion, a body in uniform motion will tend to continue travelling in a straight line unless an external force is applied to make it change direction. Therefore, when we cause a body to move in a circular path, a continuous force has to be applied to keep it changing direction and to prevent it flying off in a straight line that would be tangential to the circular path it is rotating in. This force is called *Centripetal Force* and it acts inwards towards the centre of the circular path.

According to Newton's third law, to every action there is an equal and opposite reaction. In this case, the reaction to the centripetal force is called *Centrifugal Force* and this acts in an outwards direction.

CHAPTER THREE
THEORY OF FLIGHT

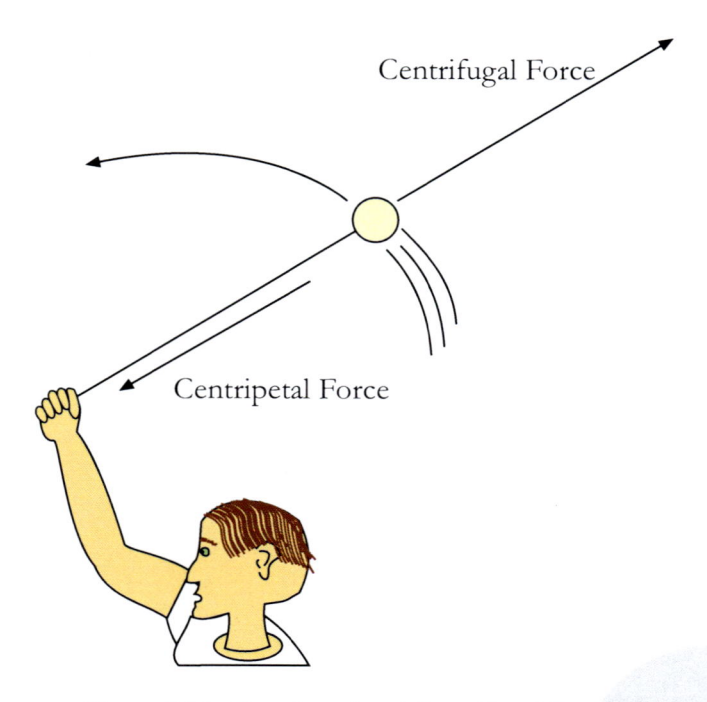

Figure 3.16 - Centripetal and Centrifugal Forces

Force is the product of mass and acceleration, so we can say that our rotating object has an inwards acceleration. This is called ***Centripetal Acceleration***. When you multiply this by the mass of the object you obtain the value of the centripetal force. There is also an outwards acceleration that together with the mass of the object gives you the reaction or centrifugal force.

Consider an object rotating clockwise at constant velocity in a circular path.

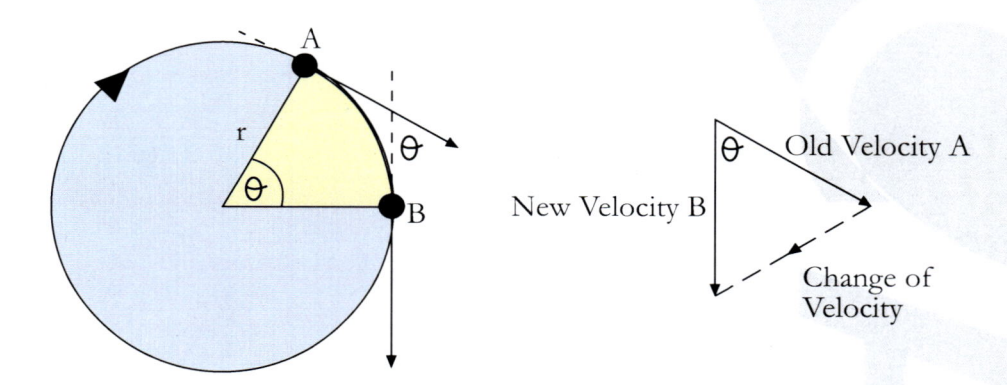

Figure 3.17 - Rotating Object

At point A, the linear velocity (v) is in the tangential direction indicated. At point B the linear velocity (v) is unchanged but the linear direction has changed as indicated. If the direction has changed then there must be a velocity involved in the change of direction from point A to point B. This is shown on the vector diagram as 'change of velocity'. If you look at the diagram again you may see that the vector representing the change would point in towards the centre of the circle. As the velocity change occurred in the time the mass moved from point A to point B then this change of velocity over time is actually acceleration – it is the centripetal acceleration.

The equations we need to determine the centripetal acceleration and force are shown below.

$$\text{Centripetal Acceleration (a)} = \frac{v^2}{r}$$

$$\text{Centripetal Force (F)} = \frac{mv^2}{r} \text{ or } \frac{Wv^2}{gr}$$

Where:
- a   linear acceleration
- F   force
- G   acceleration due to gravity
- m   mass
- r   radius of circular path
- v   linear velocity
- W   weight

Note: W/g is an expression that is exactly the same as the mass (m)

A few relationships come from these formulae:

Centripetal force is directly proportional to the mass of an object that is in circular motion. If the mass of the object is doubled at constant speed, the centripetal force will double to keep the object travelling in a circular path.

Centripetal force is inversely proportional to the radius of the circle in which an object travels. If the radius of the turn is reduced at constant speed, the centripetal force will increase to force the object away from its linear tendency into the smaller radius circular path. The directional change is greater so the force required to make that change must be greater.

Centripetal force is proportional to the square of the velocity. *If the velocity is doubled, the centripetal force rises four times. If the velocity triples, the force rises nine times.*

Aircraft Turning & Banking

As with objects whirling around on a cord, aircraft that are turning on a circular flight path will require centripetal force to hold them in the circular path. In this case, the centripetal force is created by the horizontal component of the aircraft lift. To achieve this, the aircraft has to adopt an angle of bank (θ). However, if this angle is too great, the aircraft will be pushed inwards towards the centre of the turn, if the angle is not big enough the aircraft will slide out of the turn. At the correct angle of bank, the aircraft will continue to turn on a constant radius. If you were to hold a cup of coffee on the flight deck when the aircraft is in this condition the liquid in it would remain level. If the angle of bank were too big, the coffee would spill inwards towards the centre of the turn. If the angle of bank were too small, the coffee would spill outwards away from the centre of the turn. The coffee acts as a crude turn and slip indicator.

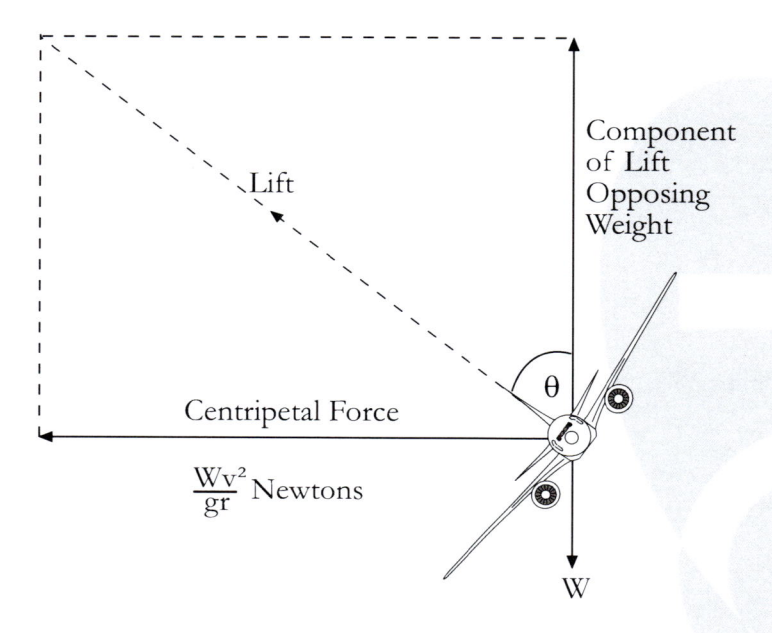

Figure 3.18 - Aircraft Turning

To determine the correct angle of bank (θ) for a given radius of **turn (r) and** airspeed (v) we can use simple trigonometry.

$$\text{Tan } \theta = \frac{\text{Opposite}}{\text{Adjacent}} = \frac{(Wv^2/gr)}{W} = \frac{v^2}{gr}$$

Note: the aircraft weight (W) cancelled out top and bottom to **give us the final** expression. Therefore, the weight of an aircraft has absolutely no effect on the angle of bank (θ), turn radius (r) or the airspeed (v). It is just the velocity and the radius of the turn that affect angle of bank.

We can try and solve a problem.

An aircraft is flying on a circular path radius 1000m at a velocity of 100m/s. What is the correct angle of bank? Assume g to be 10 m/s².

$$\tan \theta = \frac{v^2}{gr} = \frac{100 \times 100}{10 \times 1000} = 1$$

$$\theta = 45°$$

You could solve other problems by transposing the formula to find the airspeed (v) or the radius of turn (r). Also note that this is an easy question to ask on a multi-choice examination. The tangent of 45° is 1 and examiners will assume everybody knows that because it is in module 1! A rough approximation of the acceleration due to gravity is 10m/s² and that makes its use in calculation easy.

That is the physics bit of the theory of the turn. We can now look at the practical aspects of carrying out this manoeuvre.

The Level Turn

The level turn may be either a medium level turn or a steep turn. The bank angles of the former are 30° or less whilst the latter are 45° or more. Banking the aircraft produces the centripetal turning force from the horizontal component of lift. As there are no other forces to counteract it the aircraft is 'pulled' into a circular flight path. The larger the bank angle is, the greater the centripetal force will be and that will reduce the radius of the turn.

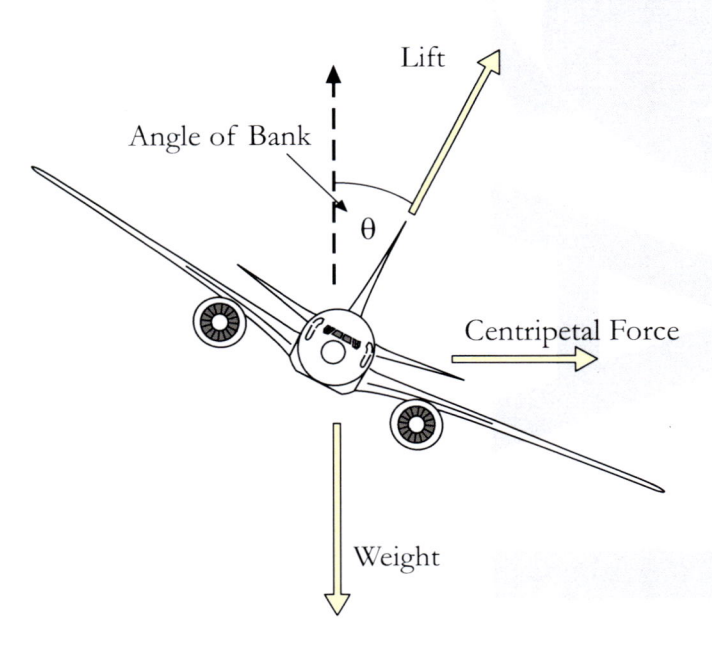

Figure 3.19 - Forces on an Aircraft in a Turn

An increase in airspeed will require an increase in the angle of bank.

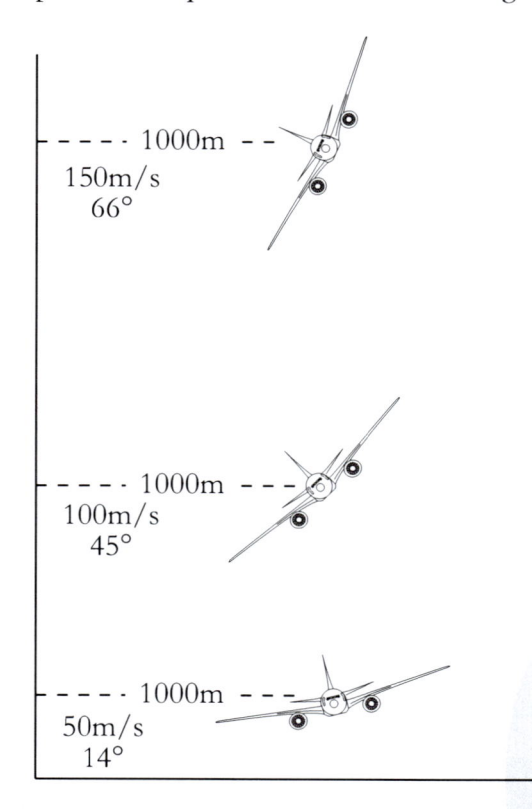

1000m
150m/s
66°

1000m
100m/s
45°

1000m
50m/s
14°

Figure 3.20 - Angles of Bank at 1000m Radius (Varying airspeed)

A decrease in the radius of the turn will require an increase in the angle of bank. The speed has a lot more effect on the angle of bank than the radius does.

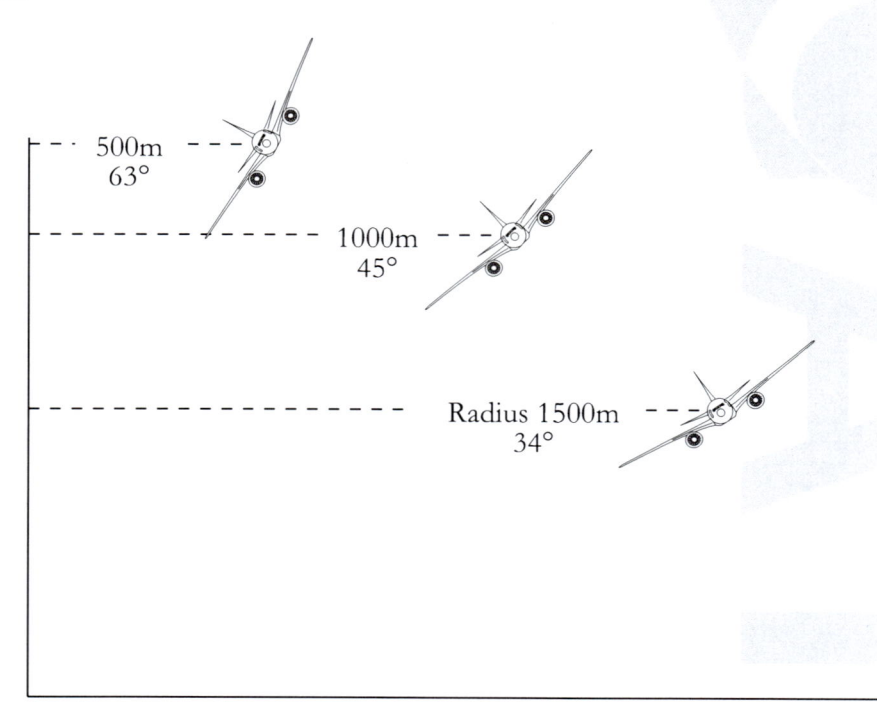

500m
63°

1000m
45°

Radius 1500m
34°

Figure 3.21 - Angles of Bank at 100 m/s airspeed (Varying Radius)

Turning tilts the lift vector away from the perpendicular and this will cause the aircraft to lose height unless the vertical component of the lift is increased. This means the Pilot must increase the lift produced by the wings. Lifting the nose of the aircraft will increase the angle of attack of the wings. This is done until the vertical component of lift balances the aircraft weight again. The attitude of an aircraft in a turn will be more nose-up than in level flight. The degree of pitch up required is correct if the aircraft does not gain or lose height in the turn.

The increase in lift required during a turn also produces an increase in induced drag that will tend to reduce the aircraft speed. The speed has to be maintained at the correct value so an increase in engine power will be required during the turn. The engine power is set to be higher in the turn than in level flight.

The basic stalling speed V_{S1} of the aircraft will increase during a turn. The greater the angle of bank the higher the stalling speed will be. The reason for this is that the wings have to be set at a higher angle of attack in a turn than they would need in straight and level flight. The increased lift increases the load factor (n) borne by the wings and thus the stalling speed increases. The aircraft weight increases under the influence of the **centrifugal force**. The increase in weight has to be balanced by an equal increase in lift. Stalling will occur when the wings cease to provide sufficient lift to balance the weight. This occurs at the stalling speed. The variation in stalling speed with loading can be calculated.

$$\text{New Stalling Speed} = \text{Normal Stalling Speed} \times \sqrt{n}$$

Where n is the load factor expressed as a multiple of g.

For example: If the load factor is 2g then: $\sqrt{n} = \sqrt{2} = 1.414$

If the normal stall speed is say 100kts then: 100kts × 1.414 = 141.4kts

At 60° angle of bank the lift on the wings will be **double** what it would be in level flight. At 70° the lift triples and at 75° it quadruples. There is a calculation for this:

$$\text{Lift} = \frac{W}{\cos \theta}$$

Note: At 60° the cosine value is 0.5 so lift will have doubled.

Figure 3.22 - Lift and Stalling Speed Changes During a Turn

The angle of bank, lift, drag, engine power and stalling speed all increase as the radius of the turn is reduced. There are a few limits to the turning manoeuvre. The wings may stall when the lift reaches a limiting value. The maximum available engine power may limit the tightness of the turn. The structural load limit may be reached.

A problem experienced during level turns is that of over-banking. During the level turn manoeuvre, the outside wing is travelling faster than the inner wing so it produces more lift. This will result in an increasing angle of bank unless the Pilot corrects it using the controls.

To maintain the turning manoeuvre the angle of bank is controlled using the ailerons. The rudder is used to maintain balance and the elevator is used to control the angle of attack and maintain height.

We should finally consider the turn manoeuvre when it is conducted during either a climb or a descent.

Climbing Turn

The forces in a climb have been examined. If the aircraft is turned during this manoeuvre, the lift will tilt and will not be sufficient to balance the weight. The result is that the climb performance will be reduced in that the rate of climb will reduce. The rate of climb for an aircraft depends on the excess thrust available to overcome drag. The reduction in lift and increase in drag during the turn will reduce the excess thrust available. The greater the angle of bank the more the rate of climb will decrease. To maintain airspeed the aircraft nose can be lowered, reducing the angle of climb.

There is a tendency for the aircraft to over bank in a climbing turn. This occurs because the wing on the outside of the turn has to travel a greater horizontal distance than the inside wing in order to gain the same height. This increases the angle of attack of the outside wing and together with the increased speed of the airflow over it this creates an additional lift that raises the wing into a steeper angle of bank.

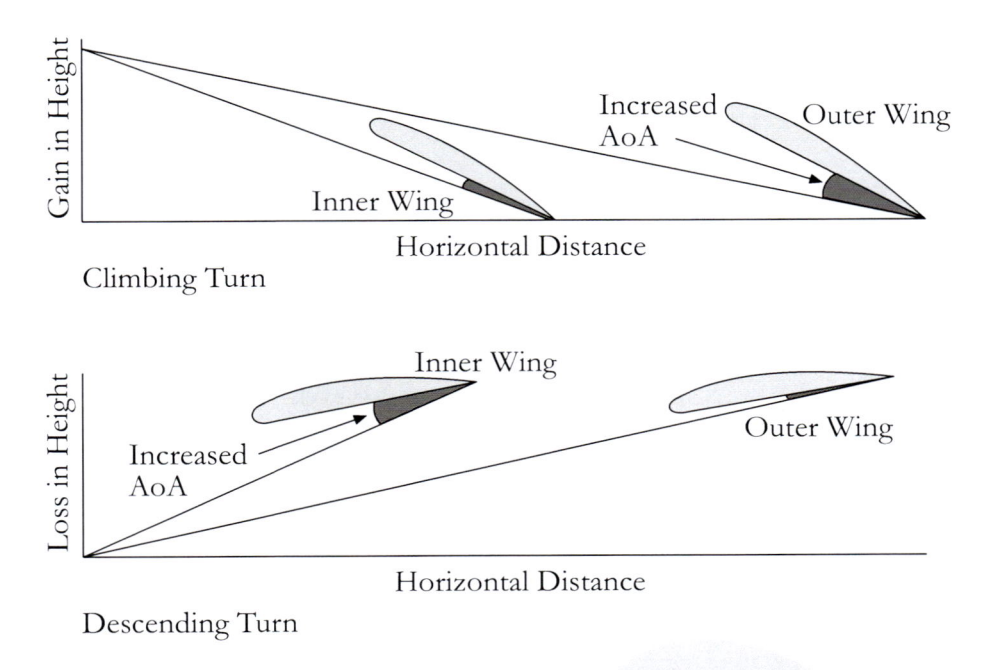

Figure 3.23 - AoA Changes During Climbing and Descending Turns

Gliding Turn

If the aircraft is turned whilst gliding, the tilt to the lift vector will reduce the effective lift required to balance the weight and the aircraft will increase its rate of descent in a steeper glide. The drag will increase and this will decrease the airspeed. In a tight turn this could bring the aircraft close to its stalling speed that has increased because of the turn. The airspeed can be maintained by lowering the aircraft nose to increase the angle of glide.

There is less of a tendency for the aircraft to over bank in a descending turn as explained below.

Descending Turn

The rate of descent can be controlled by use of the engine thrust and the pitch control. An increase in thrust and raising the nose to control airspeed will reduce the rate of descent and will reduce the angle of descent. A reduction in engine thrust and dropping the nose to control airspeed will increase the rate of descent and increase the angle of descent.

The tendency of an aircraft to over bank in a turn is less when it is in a descending turn. In this configuration the wing on the inside of the turn will have an increased angle of attack. The increased lift produced by this is counter-balanced by the additional lift produced by the outer wing travelling faster.

CHAPTER THREE
THEORY OF FLIGHT

Influence of Load Factor

We refer to multiples of g as the load factor (n) where 1g is equivalent to load factor n = 1. Objects at rest on the Earth are subject to the acceleration due to gravity only and are said to be experiencing a load of 1g, their normal weight. Additional accelerations will alter the g loading on the object and its apparent weight will change.

We have already examined why there is an apparent increase in aircraft weight during the turn manoeuvre and why this increases the stalling speed of an aircraft. We need to examine the physics lying behind the apparent weight changes and their relationship with acceleration.

You are the weight you are because the mass of your body is under the influence of the force of gravity. Force is the product of mass and acceleration. You have the mass and gravity provides the acceleration. Under normal circumstances on the surface of this planet the acceleration due to gravity is 9.81 m/s² or 32 ft/s². So, if your mass is say 70kg and you multiply it by the acceleration due to gravity your weight will be around 687 Newton (N) or about 154lb. Weight is a force, mass is not. You may have heard of the term g when applied to aircraft in flight. When you are standing still on the ground you are experiencing a force of 1g, your normal weight. This is known as load factor n = 1.

Imagine that you are now standing in an elevator. The elevator drops with an acceleration of 9.81 m/s². The floor of the elevator is falling away from under your feet at exactly the same acceleration as gravity is imposing on you. You will experience 0g as you are weightless. The load factor here is n = 0.

Now imagine that the lift stops and then accelerates upwards at 9.81 m/s². The floor of the elevator is now pushing up on your feet with the same acceleration that is trying to pull you down. You will experience 2g and you are twice your normal weight. The load factor is n = 2. These apparent weight changes can only occur during actual acceleration or deceleration. At constant velocity you will not alter from your normal weight.

Deceleration can also create g forces. Imagine you are descending at constant velocity when the elevator starts to decelerate at –9.81 m/s². You were already under the influence of gravity, now your feet are pushing against the floor of the lift with an additional force. You will experience 2g, twice your weight. Load factor n = 2. This will prevail as long as the deceleration lasts.

Now imagine an aircraft in straight and level flight. It is experiencing 1g or n = 1, its correct weight. A sudden gust of wind now increases lift and accelerates the aircraft upwards. You are now of course a passenger and you and your in-flight meal tray feel pressed into your seat. Both you and the aircraft have apparently increased weight during the vertical acceleration. The aircraft then drops with an acceleration of 9.81 m/s² and you become weightless. So does the aircraft. The vertically downward acceleration increases to 14.7 m/s². You have left your seat, your in-flight meal has left you and you are restrained only by the seat belt. You are experiencing –1/2g. Load factor n = -0.5. You are half your weight but it is now acting upwards.

Wing loading may be calculated by dividing the total weight of the aircraft by the plan form area of the wings. The loading will alter in direct relation to the load factor n.

$$\text{Wing Loading} = \frac{\text{Aircraft Total Weight}}{\text{Total Wing Area}}$$

The flight load factor represents the ratio of the lift and the weight of the aircraft. A positive load factor is where the lift acts vertically upwards. This is easily reconcilable with our examination of the g force relationship with the load factor. If the lift and the weight were equal we would have a load factor of n = +1. If the lift force is double the aircraft weight for example, it equates to a vertical up acceleration of 9.81 m/s² and that equates to the aircraft experiencing 2g or a load factor of n = +2.

$$\text{Load Factor (n)} = \frac{\text{Lift Force}}{\text{Weight}}$$

Elastic materials, for example metals, have a range when if they experience increasing tensile or compressive forces they stretch or compress elastically. If the force is removed they return to their original dimensions. They obey Hooke's law where force is proportional to extension in the elastic range.

If the forces are increased further, a limit is reached where the material gives and stretches permanently. This is the limit of the material's proportionality, known as its *proof load limit*. Any increase in force above this limit will result in permanent deformation. The material is now said to be in its plastic range.

If the force increases yet further, a point is reached where the material separates in a fracture. This is its *ultimate load limit*. Design engineers are interested in both the proof and the ultimate load limits for the structures they design. You need to know how much stress an aircraft structure can bear before it will deform. You would also need to know the stress limit where the aircraft would break apart should it be exceeded.

Having established both the proof and ultimate load limits for your structure I think you would consider having a safety margin. If the Severn Bridge were allowed an *operating load limit* that coincided exactly with its proof load limit, a sparrow landing on the roof of your car could deform the bridge! The operating load limit for the bridge is probably about a quarter of the proof load limit. A proof factor of safety of four.

Now we need to look at what stress actually is:

$$\text{Stress} = \frac{\text{Load}}{\text{Cross-Sectional Area}}$$

Pressure is Force/Area so stress is essentially mechanical pressure.

Look at the above and you will see that the only methods we can use to reduce the stress are to decrease the load or, increase the cross-sectional area. For aircraft operations, a limit to the loads we can subject the aircraft to could make it commercially unviable. So, we could increase the cross-sectional areas of all the structural components. Now the aircraft could be so heavy it will not fly, or carry passengers. There seems to be only one solution left, reduce the *factor of safety*!

Firstly, we need to establish what loads we actually expect the aircraft structure to carry. To do this the designer produces what is called a V_n diagram. This establishes an operating envelope of load factor (n) and equivalent air speed (V). Typically, the design operating limits for a large commercial aircraft are +2.5g and –0.5g.

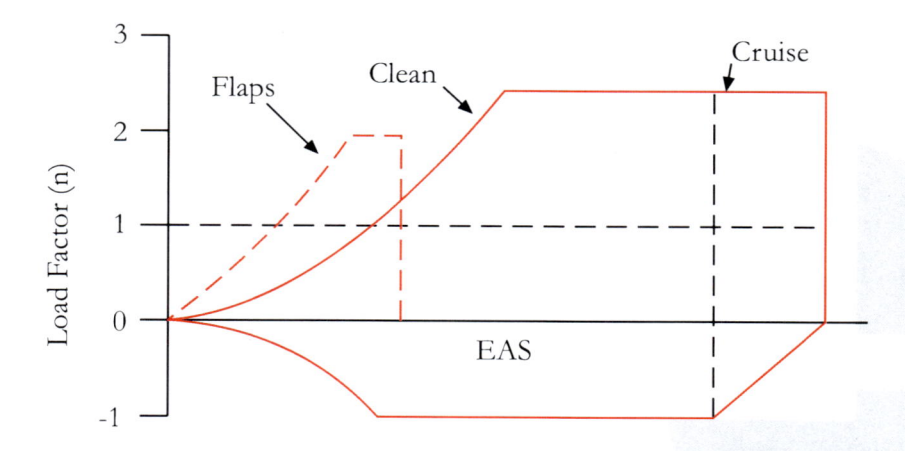

Figure 3.24 - Manoeuvring Envelope (V_n diagram)

We now have to consider the three most important load limits for aircraft.

1. The proof load limit (if exceeded the aircraft will deform)

2. The ultimate load limit (if exceeded the aircraft will break up)

3. The design operating load limits (the never to be exceeded limits during flight)

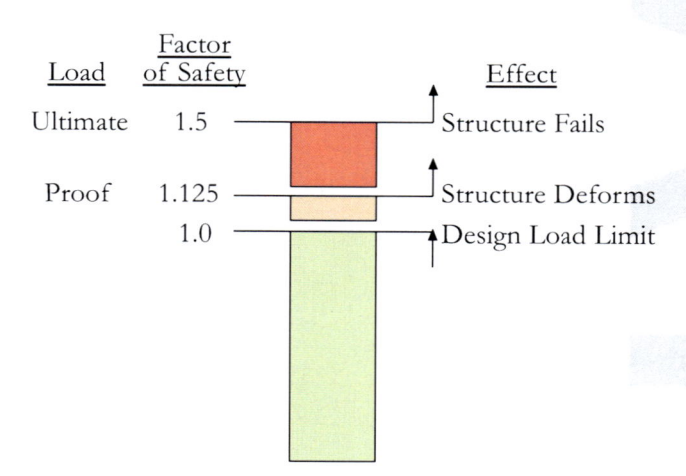

Figure 3.25 - Aircraft Structural Design Safety Factors

The design operating load limits we used in the example were +2.5g and −0.5g. These relate to the maximum upward and downward accelerations we can expose the aircraft to continually. For example if an aircraft weighing 400ton were to experience +2.5g the apparent weight would rise to 1000ton! Exposure to −0.5g would result in an apparent weight of -200ton - acting upwards!

Now for the factors of safety we are going to apply. For reasons of commercial viability, the ***proof factor of safety is 1.125 times the design operating load limits***. Not a lot! In fact the 0.125 is only there because that is the material reserve limit that is applied by the sheet metal and bar manufacturers. So, if we exceed +2.5g or −0.5g the aircraft will deform.

The ***ultimate factor of safety is 1.5 times the design operating load limits***. This gives us limits of +3.75g and −0.75 after which the aircraft breaks up! Our 400ton aircraft would have an apparent weight of +1500ton at 3.75g. At this loading, the wing tips would have flexed up about twenty-four feet! Failure would normally appear initially as severe buckling of the top inner wing skins followed by the explosive fracture of both wings. Not a pretty sight! And your in-flight meal has gone somewhere else too!

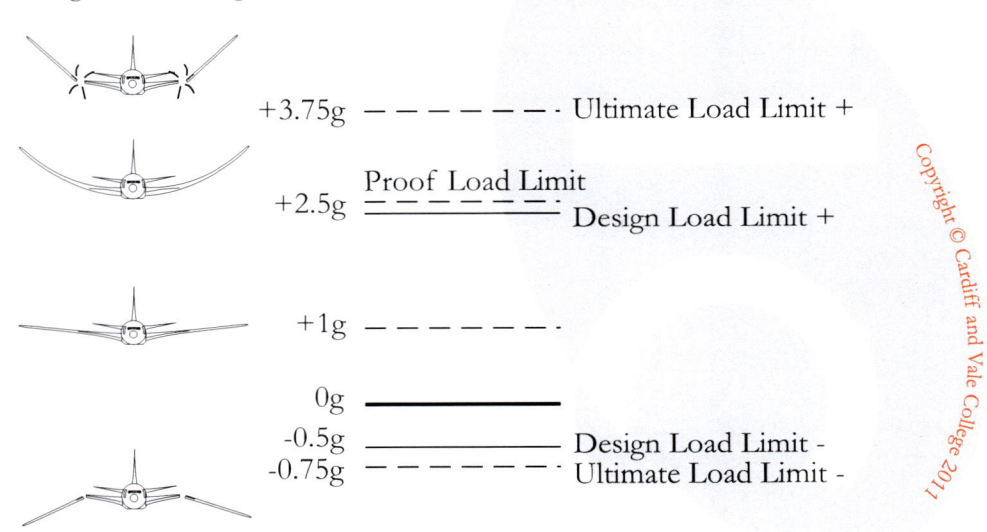

Figure 3.26 - Structural Load Limits

Stall

To remind you, stalling occurs when the streamline flow over the wings becomes turbulent and separates from the top of the wing. This occurs at the critical angle of attack, around 15°. Stalling will occur whenever the critical angle of attack is exceeded regardless of the airspeed. The onset of a stall is signalled by turbulent air streaming back off the wings and buffeting the tail producing a shuddering motion that is felt through the aircraft structure. This occurs at the pre-stall or incipient stall period. As stalling occurs, the lift will decrease causing the aircraft to sink, the centre of pressure will move rearwards and the nose will drop.

The basic stalling speed V_{S1} of an aircraft is normally specified as the speed where an aircraft would normally reach the fixed stalling angle of attack at its maximum weight with the flaps retracted. You must remember that stalling always occurs at the same angle of attack not at the same airspeed. The Pilot may on occasions have to increase the angle of attack earlier to increase lift. Instances where this may occur could be during a turn, pulling out of a dive, landing or flying slowly. In each case the speed at which the critical angle is reached will be different.

Effect of Altitude on Stalling Speed

The stalling speed of an aircraft will increase with an increase in altitude due to the lower air density producing less lift. For a given airspeed the angle of attack required to maintain lift at high altitude will be higher than that required at low altitude. Thus, when airspeed is reduced the wings will reach the critical angle at a higher speed.

Effect of Weight on Stalling Speed

A heavy aircraft requires a proportionately higher lift to balance its weight. If the weight of an aircraft is increased then a relatively higher angle of attack is required for a given airspeed. Again, the speed at which the aircraft will reach the stall angle of attack will increase.

Effect of Flaps & Slats on Stalling Angle and Speed

Extension of the trailing edge flaps will reduce the stalling angle of the wing and also reduce the stalling speed. Leading edge slat extension will increase the stalling angle.

Effect of the Load Factor on Stalling Speed

In straight and level flight the load factor (n) is 1. Turning, pulling out of a dive or flying through turbulence will produce acceleration loads that will increase the load factor.

The problem with an increase in the load factor is that it increases the aircraft's apparent weight. The lift produced by the wings has to be increased to balance any increase in apparent weight if height is to be maintained. This demands an increase in the angle of attack so the stalling speed of the aircraft will increase. If the load factor is too high the aircraft could be forced into a stall condition before the lift could even reach a balance with weight. This could occur in a very tight turn for example.

You will have examined the calculation for revised stalling speeds when related to variations in the load factor (n) during the section on turning. The calculation can be used whenever the load factor changes, not just in turns. If an aircraft having a normal stalling speed of say 100kts experienced a 3g loading when pulling out of a dive for example, the revised stalling speed during this time would be: $100 \times \sqrt{3} = 173$kts. You can appreciate the concern this will create, as that is a big rise in the stalling speed.

Effect of Engine Power on Stalling Speed

As a general rule an increase in engine power will reduce the stalling speed of an aircraft. Propeller driven aircraft will benefit from the increased air velocity from the propeller slipstream passing over the region of wing in its path. If this is the wing root then the stall may well be delayed if the aircraft normally stalls at the wing roots first. For both propeller and turbojet driven aircraft, the nose up attitude of the aircraft during an incipient stall will produce a small component of vertical thrust to assist lift. This will delay the need to increase the angle of attack. Following a stall the aircraft will recover quicker if engine power is available.

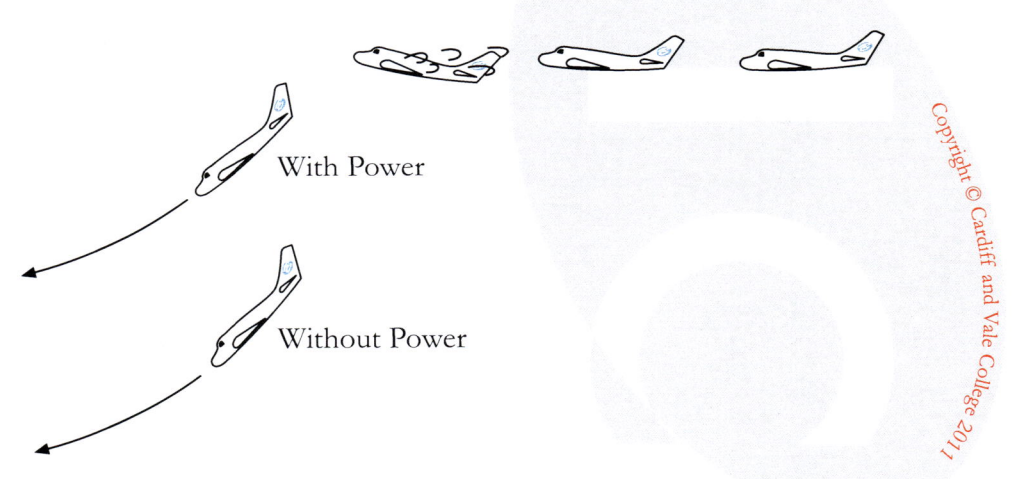

Figure 3.27 - Stall Recovery With and Without Engine Power

Effect of Icing on Stalling Speed

Ice can significantly alter the contours of an aerofoil and add to the weight of the aircraft. This will not only increase the stalling speed but can even reduce the value of the critical angle of attack where stall will occur increasing the stalling speed yet again. Frost and rime ice will increase skin friction and thus profile drag, slowing the aircraft.

Effect of Centre of Gravity Position on Stalling Speed

If the centre of gravity of the aircraft is too far forward the aircraft will develop a nose down attitude. This will be corrected by an increase in force acting downwards on the tail plane. Because this is a downward acting force it effectively increases the apparent weight of the aircraft. This will increase the stalling speed.

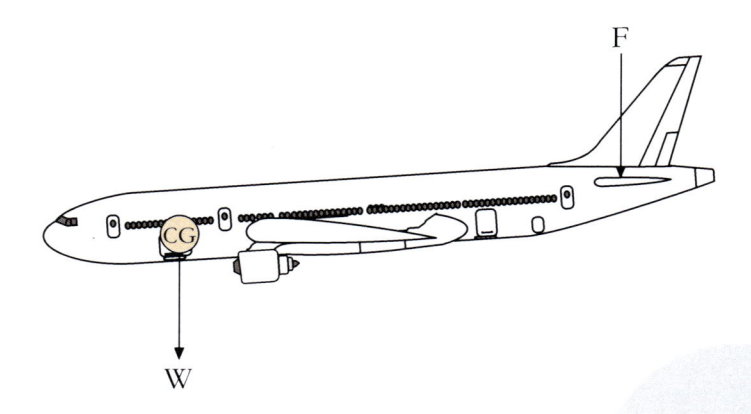

Figure 3.28 - Effect of Forward C of G

If the centre of gravity is positioned too far to the rear the aircraft will adopt a nose up attitude. The tail plane will exert an upward force in an attempt to correct this. This should reduce the stalling speed but as airspeed reduces, the correcting force also reduces and the return of a nose up tendency can bring the aircraft closer to the stall. During a stall the centre of pressure moving rearwards may be able to counteract this. It is a pretty unstable situation best avoided. If the aircraft has sweptback wings, the stall may well originate at the wing tips. This has the effect of moving the centre of pressure forwards and lifting the nose still further. A situation could be reached where recovery is impossible.

If the centre of pressure is too close to the centre of gravity, the aircraft may not be able to drop its nose in the stall and it will merely sink and that will put the aircraft further into the stall because the angle of attack will increase. If the centre of pressure is at a distance from the centre of gravity during the stall it will exert a nose down turning moment around the C of G and this will assist in recovery.

Effect of Aileron use

If an aircraft is near the stalling angle of attack rolling the aircraft can put the down-going wing into a stall. The down-going wing experiences an increase in angle of attack that will stall the wing causing it to drop further. This puts the wing deeper into the stall. Because the up-going wing remains un-stalled a spin could be a real possibility.

Wing Tip Stall

Sweptback wings are prone to wing tip stall. The aircraft also adopts a nose up attitude during slow speed approaches. Should the angle of attack reach a position where incipient tip stall is imminent there are risks. If one tip stalls before the opposite tip, the aircraft will roll in the direction of the stalled tip. This will increase the angle of attack of the dropping wing putting the tip deeper into the stall and possibly causing the stall to spread inboard. Any attempt to correct the roll by deflecting the ailerons will be ineffective due to the aileron being in the stalled tip region. The aircraft could then sideslip into the ground.

Stall Strips

Whilst rectangular wings stall at the wing roots first, tapered and swept back wings have a tendency to stall at the wing tips first. To encourage the stall to occur at the wing roots on these latter configurations the stall strip is frequently used. This is a triangular sectioned strip fitted along the inboard leading edges of the wings. At high angles of attack approaching the critical angle, the strips encourage the airflow passing over the leading edges in the root areas to break away. This initiates stalling in these areas before the tips reach incipient stall.

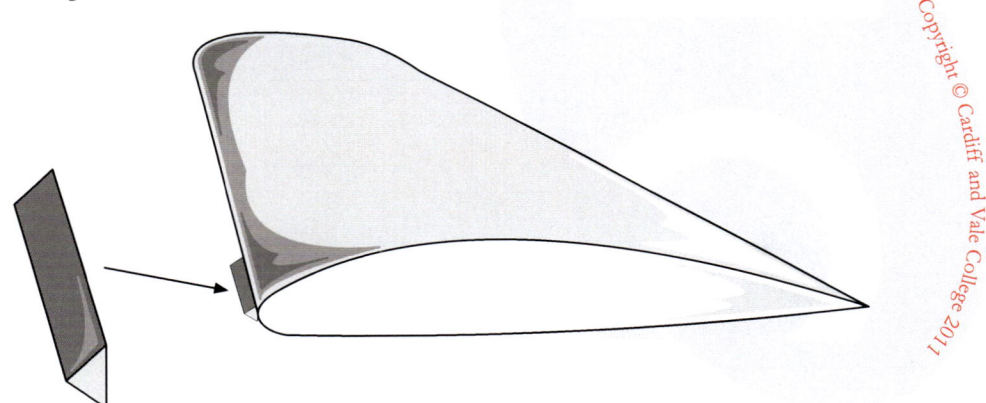

Figure 3.29 - Stall Strip

Stall Sensing

If the configuration of the wing is such that the wing initiates stall at its root then the separated and turbulent airflow will buffet the tail of the aircraft giving a clear warning of incipient stall. A simple mechanical stall sensor may be used. This may consist of a simple hinged flap fitted just under the leading edge of the wing near the root. At a high angle of attack coincident with the onset of stall the airflow under the leading edge will attain a direction that is able to move the flap that in turn operates a micro-switch to relay the warning. More sophisticated sensors fitted on the fuselage employ a flying aerofoil vane that is mounted on the rotational axis of a potentiometer.

Deep Stall

This type of stall affects those aircraft that have a flying horizontal stabiliser or tail plane positioned on top of the vertical stabiliser or fin. The stabiliser is positioned to permit the rear mounting of the main engines. Stall recovery requires that the pilot should gently ease the control column forward to push the aircraft nose down. This action reduces the angle of attack and puts the aircraft into a descent where airspeed and thus lift can be recovered.

The correctly trimmed aircraft will tend to drop its nose naturally during the stall so the pilot is just encouraging it. Both the natural tendency and the pilot's control actions require that the tail plane is fully functional and that the elevator is effective. The turbulence streaming back off the wings would normally pass well above the tail plane because of the pitched up attitude of the aircraft during a stall. However, when the tail plane is positioned on top of the fin an excessively high pitch up attitude will put it right in the path of the turbulence rendering it ineffective. This situation is called deep stall, super-stall or locked in stall.

Figure 3.30 - Deep Stall

The aircraft will not drop by the nose and any crew control input to encourage this is ineffective. The aircraft will sink and this causes the **angle of attack** to increase further and put the wings deeper into the stall. The **aircraft will** continue to sink until ground contact is made. Hence the description **locked in** stall. The pitch up angle where deep stall becomes a risk is very high in the order of 30° and over.

High Speed Stall

We have already examined this in the last chapter. This type of stall occurs when the aircraft reaches a mach number M_{CRIT} where the airflow over any curved surface on the aircraft reaches Mach one. This is usually over the point of maximum curvature on the top of the wings. When this occurs, a normal shock wave forms along the span of the wings. This causes the **airflow passing** through it to decelerate creating a strong adverse pressure **gradient and airflow** separation. This stall is often referred to as a shock stall. As altitude increases the aircraft will reach this critical Mach number M_{CRIT} earlier. This is due to the fact that the value of the speed of sound in air reduces with the reducing air temperature at altitude.

The speed at which the low speed stall occurs rises with altitude and, as we have already examined, the aircraft can become trapped between the low and the high speed stalls. The next time you hear of an aircraft that somehow miraculously dropped four thousand feet for no apparent reason you might like to reflect on whether the crew were awake at the time. If the aircraft altitude creeps up unnoticed the risk of this increases. There is an alarm fitted to the Machmeter to alert crews of a close approach to the critical Mach number. Once trapped, any variation in airspeed triggers one form of stall or the other!

Lift Augmentation

The secondary flight controls on an aircraft are chiefly used to either augment or destroy lift. The controls used to augment lift are: Trailing edge flaps, leading edge slats and slots and leading edge flaps. The controls used to destroy lift and in some cases increase drag are: Flight spoilers, ground spoilers and speed brakes. We will deal with each of these in this section.

Trailing Edge Flaps

Extending the trailing edge flaps alters the shape of the wing to create an increase in lift and a bigger increase in drag. The lift coefficient is affected by the camber of the wing. As the camber is increased the maximum obtainable value for the coefficient of lift will increase. This is often supplemented due to an increase in the wing plan form area. This increases the lift but the lift/drag ratio will be reduced because of the proportionally greater increase in drag.

The stalling speed reduces as the flaps extend enabling flight at low airspeeds where the flaps will still maintain sufficient lift. Extension of the trailing edge flaps will however decrease the stall angle of attack. The lower value for the stall angle results from the change in the aerofoil shape altering the effective angle of attack. The further the flaps are extended, the lower the stall angle becomes and the lower the stalling speed becomes.

As the trailing edge flaps extend, the changing shape of the wing and the change in aerodynamic forces moves the centre of pressure rearwards tending to pitch the aircraft nose down. The increased downwash caused by flap extension reduces the tail plane angle of attack and this exerts an additional nose down tendency.

The maximum value for the coefficient of lift obtained by using the trailing edge flaps may be limited by flow separation on the flap upper surface. Some configurations of flap incorporate slots as a means to control boundary layer separation at high angles of attack. However, the angle of attack at which the maximum value of C_L can be obtained with flaps is less than that for a clean aerofoil.

The increase in lift imposes increased stress on the airframe so there is a maximum airspeed V_{FE} above which the flaps may not be selected. Many aircraft have safety systems fitted that either prevent the selection of the flaps above V_{FE} or trigger an automatic retraction should this speed be exceeded. There are a number of different configurations of flap. The choice of which type is fitted depends on the aircraft design.

We will examine each of these configurations in turn.

The Plain Flap

Sometimes referred to as the 'camber flap' this design forms an integral part of the trailing edge of the wing. When lowered it alters the camber of the wing but not the area. When used, the plain flap increases the lift by as much as 50% at 12° angle of attack. The design is limited in regards to maximum camber change because of the risk of flow separation on the top of the wing in the area of the flap. This flap creates a lot of drag and exerts a marked nose down pitching moment due to the rearward movement of the centre of pressure.

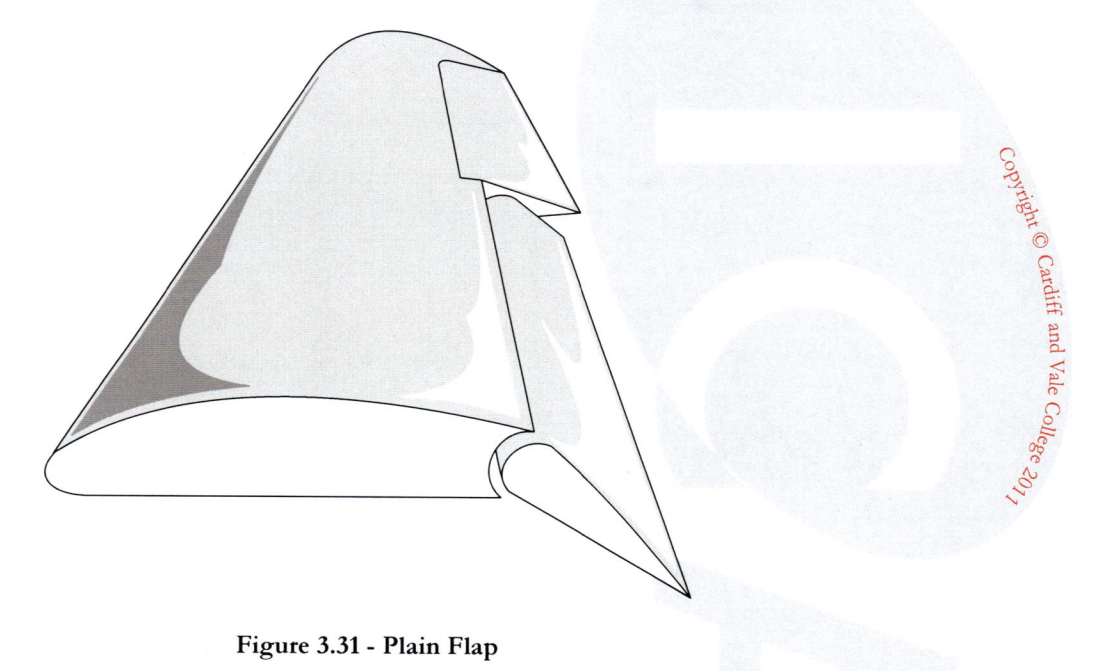

Figure 3.31 - Plain Flap

The Split Flap

The flap is a hinged panel on the underside of the wing trailing edge. When the flap is lowered it increases the wing camber by altering the under-wing profile but does not alter the shape of the top of the wing or the wing area. This design avoids the risk of flow separation over the wing top surface. The lift is increased by up to 60% at 14° angle of attack. The flap creates more drag than the plain flap and gives a marked nose down pitching moment as the centre of pressure moves rearwards.

Figure 3.32 - Split Flap

The Zap Flap

Similar to the split flap except that the flap hinge travels rearward when lowered. This flap increases the effective area of the wing as well as its camber without changing the shape of the top surface. Like the split flap there is little risk of flow separation on top of the wing. The flap creates a lot of drag and exerts a nose down pitching moment because of the rearward travel of the centre of pressure. Lift is augmented by as much as 90% at 13° angle of attack.

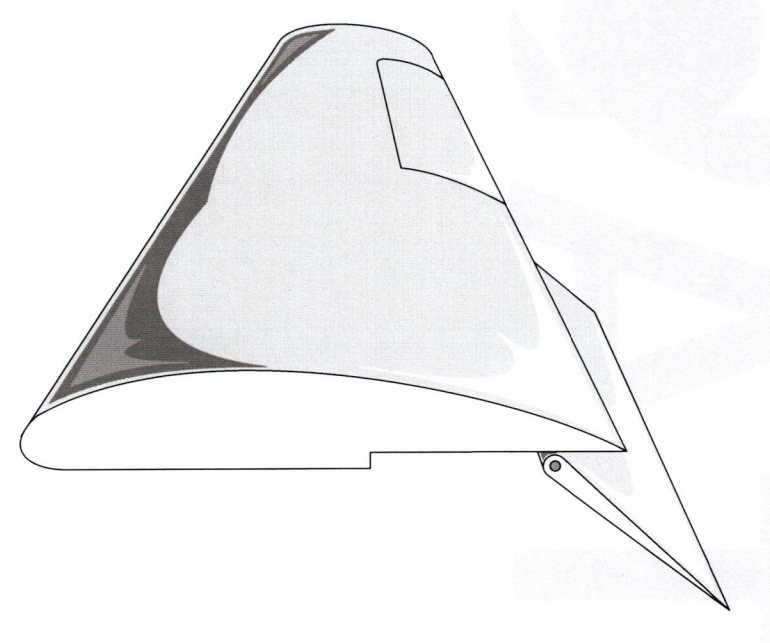

Figure 3.33 - Zap Flap

The Fowler Flap

This flap forms part of the underside of the wing trailing edge but slides rearwards and then downwards as it is extended. This design mainly increases the area of the wing as it begins to extend rearwards and then mainly the camber as it lowers towards full extension. The Fowler flap is considered to be the most effective design for producing lift but the mechanism used to operate it is complex and adds weight. The increase in lift is up to 90% at 15° angle of attack. Extension causes a nose down pitching moment due to rearward movement of the centre of pressure.

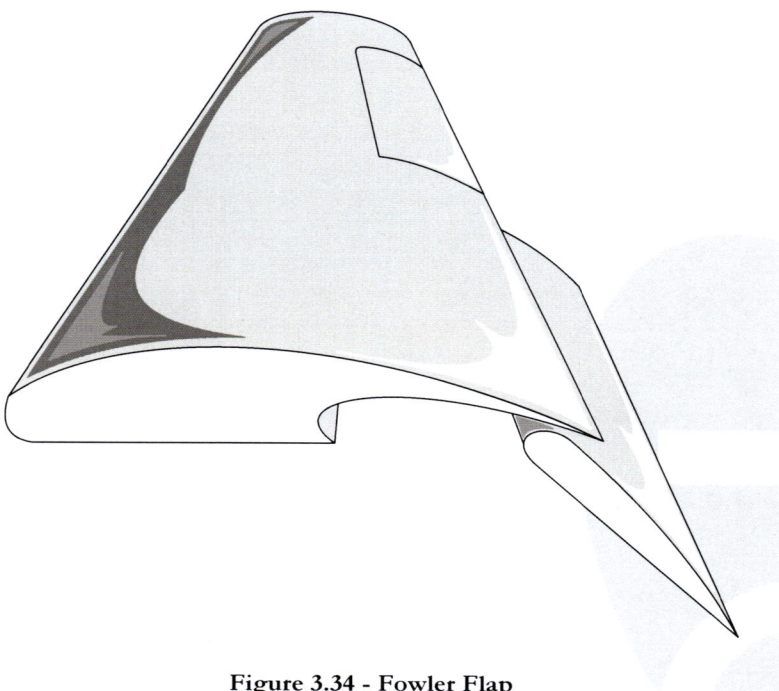

Figure 3.34 - Fowler Flap

The Slotted Flap

This flap is designed so that when lowered it creates a gap between the leading edge of the flap and the trailing edge of the wing forming a slot with a convergent section. This permits air to pass from the high-pressure region under the wing and accelerate through the convergent slot to mix with and energise the airflow passing over the flap. The objective is to control the boundary layer and prevent it separating after it has slowed in the adverse pressure gradient towards the rear of the top surface of the wing. By introducing air that has been accelerated through the slot the layer is re-energised and will tend to remain laminar over the top of the flap. That is the idea anyway. Extension of the flap increases the camber of the wing but not the area. There is not as much drag created as with previous configurations and the lift increases by up to 65% at 16° angle of attack. It gives a nose down pitching moment. This configuration of flap has been produced with up to two additional fixed slots to give double and triple slotted versions for improved boundary layer control. The double slotted flap augments lift by up to 70% at 18° angle of attack.

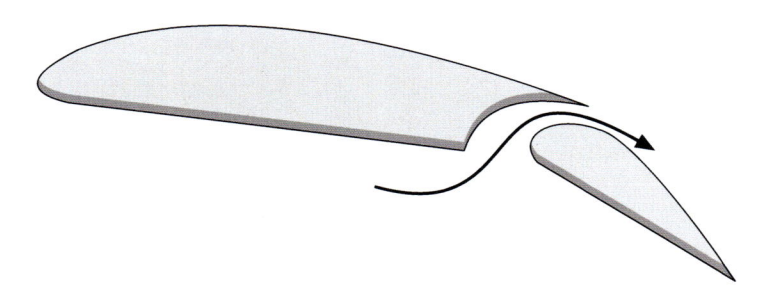

Figure 3.35 - Slotted Flap

The Slotted Fowler Flap

This design incorporates the slotted configuration into the Fowler flap design. In its simplest form the flap extends to form a single slot with the wing trailing edge. In improvements to the design the flap may be constructed in two or even three sections to give either fore flap and rear flap sections or fore flap, mid-flap and rear flap sections. This produces **double or triple** slotted versions of the flap. Both increase camber and area.

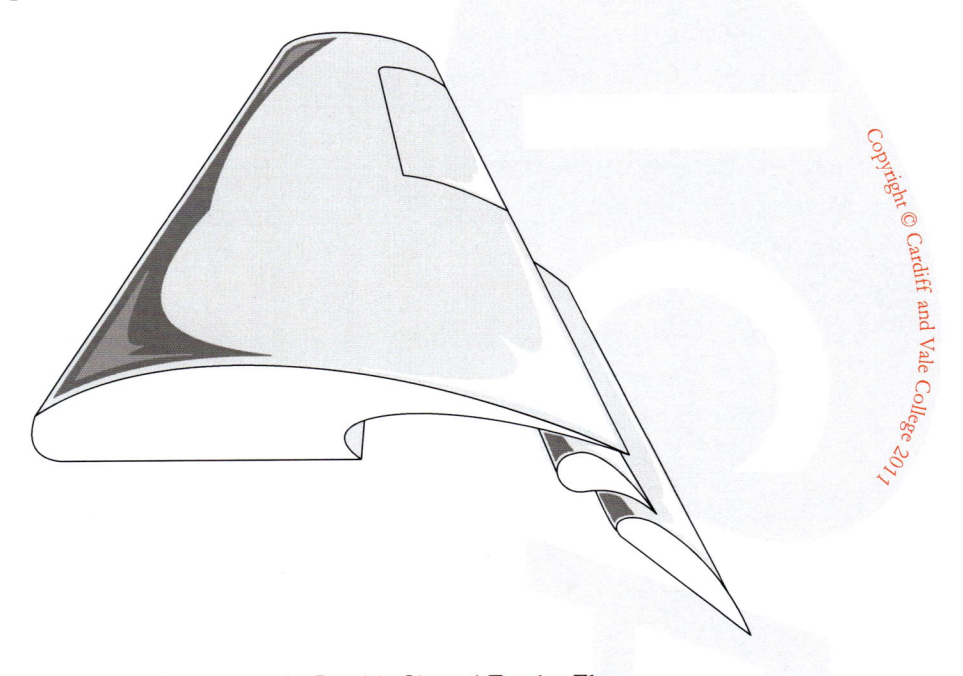

Figure 3.36 - Double Slotted Fowler Flap

The flap sections are stowed in the closed position and separate to form the required slotted configuration when extended.

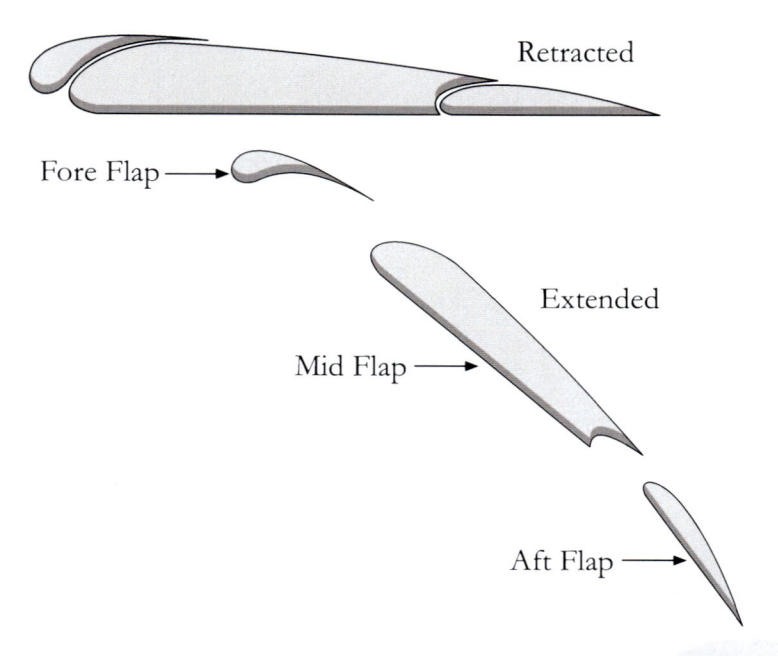

Figure 3.37 - Triple Slotted Fowler Flap

There are several flap extension selections. The takeoff selection can be made to give a varying number of degrees of extension depending on the ambient conditions at the airfield prior to takeoff and the additional lift required. These selections increase mainly area with some increase in camber to achieve the required lift for the lowest drag penalty.

Too much drag would mean the aircraft would just take longer to reach the required speed on the runway. The highest possible lift/drag ratio is the object at take-off. The landing selection would give full deployment to take advantage of the air braking action created by the increased drag. This selection drastically increases the camber and reduces the lift/drag ratio.

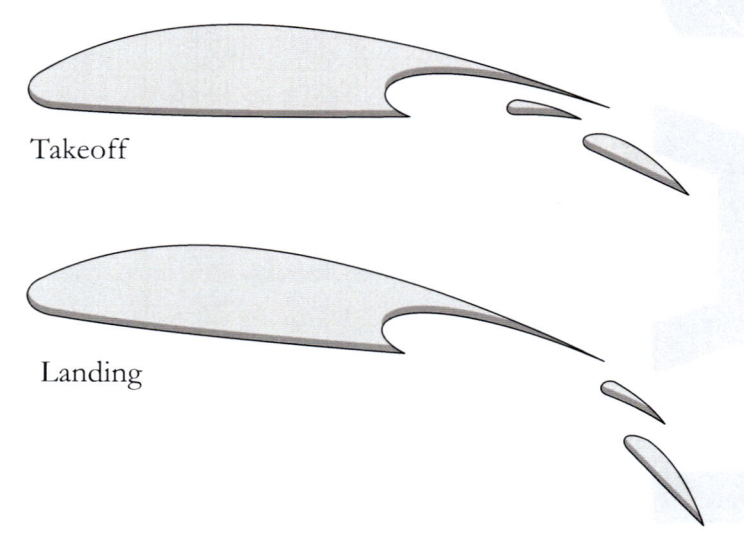

Figure 3.38 - Flap Positions

The single slotted Fowler flap will give an increase in lift up to 90% at 15° angle of attack. The double slotted Fowler flap will give an increase in lift of up to 100% at 20° angle of attack.

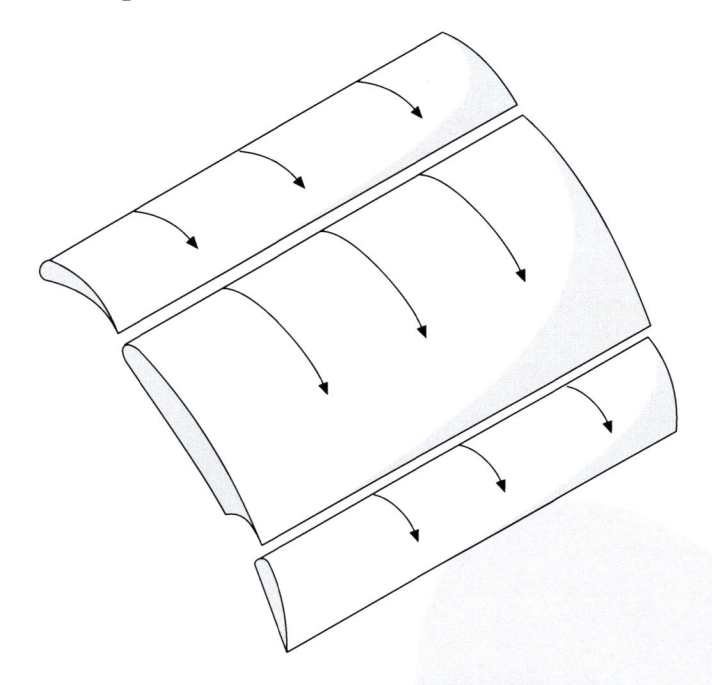

Figure 3.39 - Re-energising the Boundary Layer

Pitching Moment

All the trailing edge flaps we have examined produce a nose down pitching moment. On extension the lift increases and the aircraft can rise rapidly unless corrected. The re-distribution of the wing loading causes the centre of pressure to move rearwards producing the nose down moment. If the flaps were solely inboard flaps the centre of pressure of the wings would move inboard as well as rearwards.

Leading Edge Devices

These are used to either increase the camber of a wing to increase the obtainable coefficient of lift at low forward airspeeds or, to delay boundary layer separation over the upper wing surface at high angles of attack allowing the wing to exceed the basic stall angle and reach a higher lift coefficient C_{LMAX}. The need for these devices is brought about by the poor lift performance normally experienced by the high-speed aerofoil at low forward airspeeds.

Slots

Slots are used to control the boundary layer to prevent its separation at high angles of attack. As the angle of attack increases the lower ends of the slots are presented to the air flowing under the wing. As the pressure of this air is higher than that over the wing, the air accelerates through the convergent section formed by the slots to mix with and energise the boundary layer over the wing. Delaying boundary layer separation in this way allows high angles of attack to be maintained at low forward airspeeds. A slotted wing arrangement will augment lift by as much as 40% at 20° angle of attack. Slots are often positioned in front of the ailerons on swept wing aircraft. This helps maintain the function of the ailerons at high angles of attack where tip stalling is a risk.

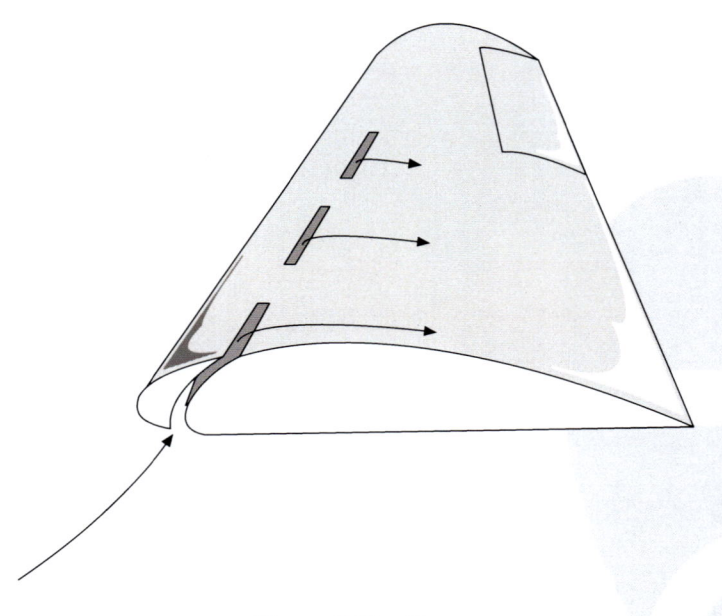

Figure 3.40 – Slots

Slats

Placing an auxiliary cambered aerofoil called a slat in front of the main aerofoil so that a convergent gap is formed between the two will delay flow separation and increase the stalling angle of the wing. Used in conjunction with trailing edge flaps this counteracts the reduction in the stall angle imposed by the flaps. The lift obtained at any given angle of attack does not vary much but the angle of attack for stall is significantly increased allowing for lift at low forward airspeeds where the wings can still achieve a higher lift coefficient C_{LMAX}. Slats can be fixed, variable or automatic. Extension of the slats will move the centre of pressure forwards giving a nose up pitch moment. This action tends to counteract the nose down moment created by the extension of the trailing edge flaps.

Fixed Slats

The fixed slat controls the boundary layer at high angles of attack by allowing high-pressure air from the underside of the wing to accelerate through the convergent gap formed between the slat and the wing leading edge and mix with and re-energise the upper wing surface boundary layer. This delays separation of the boundary layer until a very high angle of attack. The wing's stalling angle is increased. The fixed slat does increase drag in the higher airspeed range.

Movable Slats

The movable or variable slat forms a flush part of the leading edge when stowed. On selection it will move forwards to form a convergent section gap along the leading edge through which higher pressure air from below the leading edge will accelerate to energise the upper surface boundary layer and prevent its separation at high angles of attack. This allows the wing to achieve lift at higher angles of attack than it could otherwise do. The stalling angle is increased. There are normally three possible selections for these slats, retracted, extended for take-off and fully extended for landing. The latter two give a nose up pitching moment.

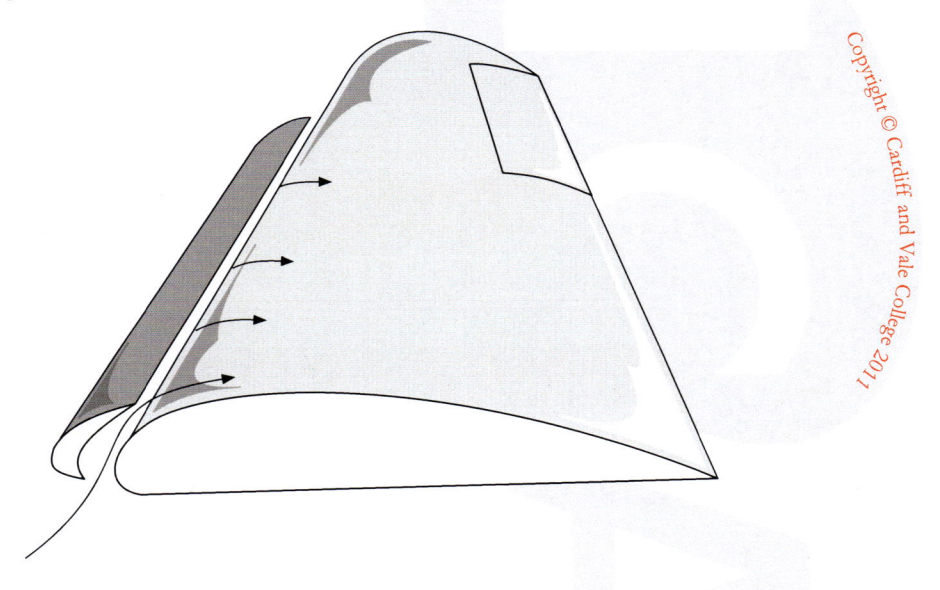

Figure 3.41 - Slat

Later configurations of the slat extend forwards and down to also increase the camber and area of the wing to further raise the coefficient of lift whilst preventing the risk of separation. Many aircraft are fitted with *Auto-slat* stall protection systems that will automatically deploy the leading edge slats when a warning of an incipient stall is received.

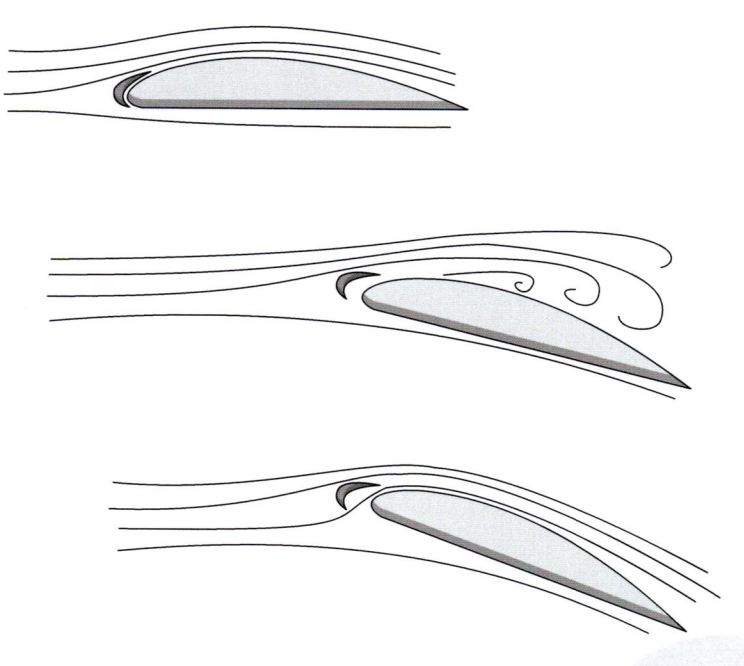

Figure 3.42 - Effect of Automatic Slat Deployment

Earlier forms of the automatic slat depended on the low pressure generated near the leading edge at high angles of attack for extension. The forward and upward suction existing near the leading edge at high angles of attack allowed spring action to extend the slat.

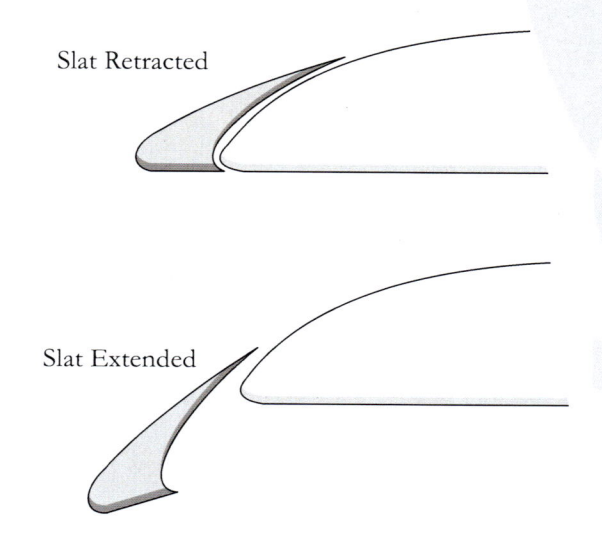

Figure 3.43 - Leading Edge Slat Extension/Retraction

A fixed slat arrangement will augment the lift by as much as 50% at 20° angle of attack. The movable or variable slat augments lift up to 60% at 22° angle of attack. A disadvantage of slats is that their successful operation depends on the aircraft adopting a high angle of attack during takeoff and landing. As this tends to be a feature of high speed swept wing aircraft operation it is not really a drawback for these aircraft types.

Leading Edge Flaps

Often referred to as the 'Krueger' flap these are installed at the inner wing leading edges. They are simply nose flaps that are hinged about the leading edge. Their prime objective is to increase lift at low forward airspeeds especially during takeoff and landing on aircraft that have high-speed configuration aerofoils. They increase the camber of the inner wings when deployed.

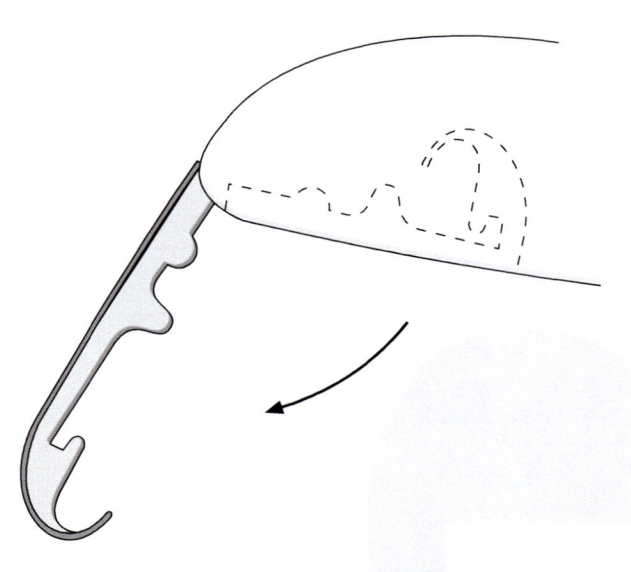

Figure 3.44 - Leading Edge Flap

When lowered, the leading edge flaps will give an increase in lift up to 50% at 25° angle of attack. Later versions of the nose flap incorporate a camber changing mechanism to produce a flap that creates a curvature on deployment.

Leading Edge Droop

Another version of the leading edge camber-changing device is the droop leading edge. On selection the whole of the leading edge section cants down to increase the curvature and thus the camber of the wing. This device works in conjunction with the trailing edge flaps.

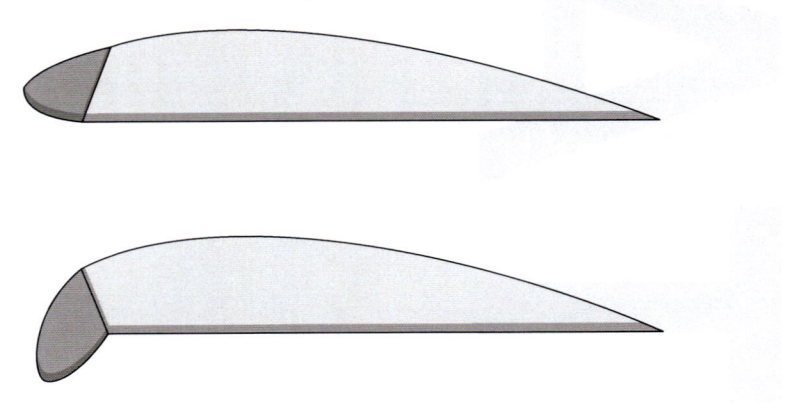

Figure 3.45 - Leading Edge Droop

Flaps & Slats

Most modern commercial aircraft embody both leading and trailing edge devices. The leading edge slats and flaps are synchronised to operate in conjunction with the trailing edge flap selections. The leading edge slats normally have three positions, retracted, extend for takeoff and full extend for landing. The trailing edge flaps have a number of extend positions for takeoff and full extend for landing.

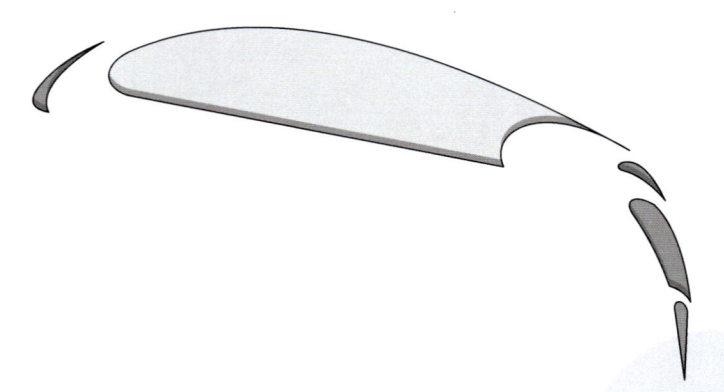

Figure 3.46 - Triple Slotted Flaps with Leading Edge Slats

When these two devices are extended together the opposite pitching moments exerted by each tend to cancel each other out. The slat and double or triple slotted flap arrangement gives up to a 120% increase in lift at an angle of attack of 28°.

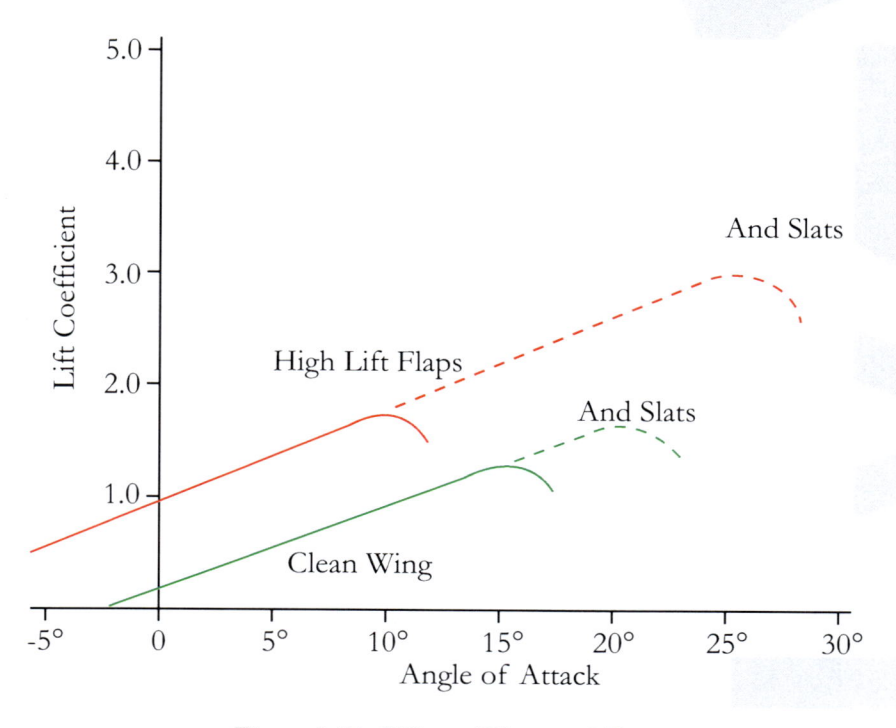

Figure 3.47 - Effect of Flaps and Slats

The Blown Flap

These have been used on a limited number of high-speed aircraft types. The system incorporates high-pressure air bled from the engine compressors being blown through nozzles into the boundary layer over the flaps. This re-energises the layer, produces a smoother and thinner laminar layer and prevents separation. The major disadvantage with the system is that the aircraft may be severely limited in its low speed operation should engine failure occur.

The Jet Flap

This is an experimental device that incorporates a sheet of high-pressure air being blown over the rear upper surface of the wings that is finally deflected downwards off the back of the wing trailing edges. This controls the boundary layer and produces very high lift coefficients. The downward efflux of high-pressure air also produces a reaction that complements the lift. A strong nose down pitching moment is experienced with the operation of the jet flap. Again, if the power source fails, the aircraft becomes limited at low speeds.

Use of Flaps for Take-off

Prior to take-off the flaps are extended to a position that will produce the required lift with as low an increase in drag as is possible. If too much drag were created the aircraft would take proportionally longer to reach the required lift-off speed on the runway. Therefore, gaining lift for the highest lift/drag ratio obtainable is the aim.

Use of Flaps in the Air

When trailing edge flaps are selected to extend, the airspeed must be below the safe speed (V_{FE}). As the flaps extend the aircraft pitch attitude changes to nose down. The effective angle of attack increases. The drag increases and at a constant airspeed the glide path will become steeper as a result. The lift drag ratio decreases. The stalling speed is reduced and the stalling angle will be reduced. Full flap extension is reserved for approach and landing where the big increase in drag becomes useful in reducing the aircraft speed and the length of the landing run.

When the flaps are raised, the lift will reduce and the aircraft will sink unless the angle of attack is increased. The flaps must only be raised when the airspeed is higher than the clean wing stalling speed. The aircraft pitch attitude will alter to nose up requiring trim and this will be added to if the engine power is increased. Flaps are rarely retracted during approach unless the landing is aborted and a go-around manoeuvre is conducted.

Spoilers

Our examination of lift augmentation would not be complete unless we looked at the device that is installed to destroy lift. The spoiler is a hinged panel that is fitted on the top rear surface of the wings. When the spoiler is deflected up into the air stream it causes the adverse pressure gradient ahead of it to increase and lift in that area will reduce at a value dependent on the angle of spoiler deflection. Drag will also increase as the spoiler increases the frontal cross-sectional area of the wing.

Spoilers are normally fitted in groups and are divided into those spoilers used in flight and those used solely on the ground to destroy the lift on the wings after landing.

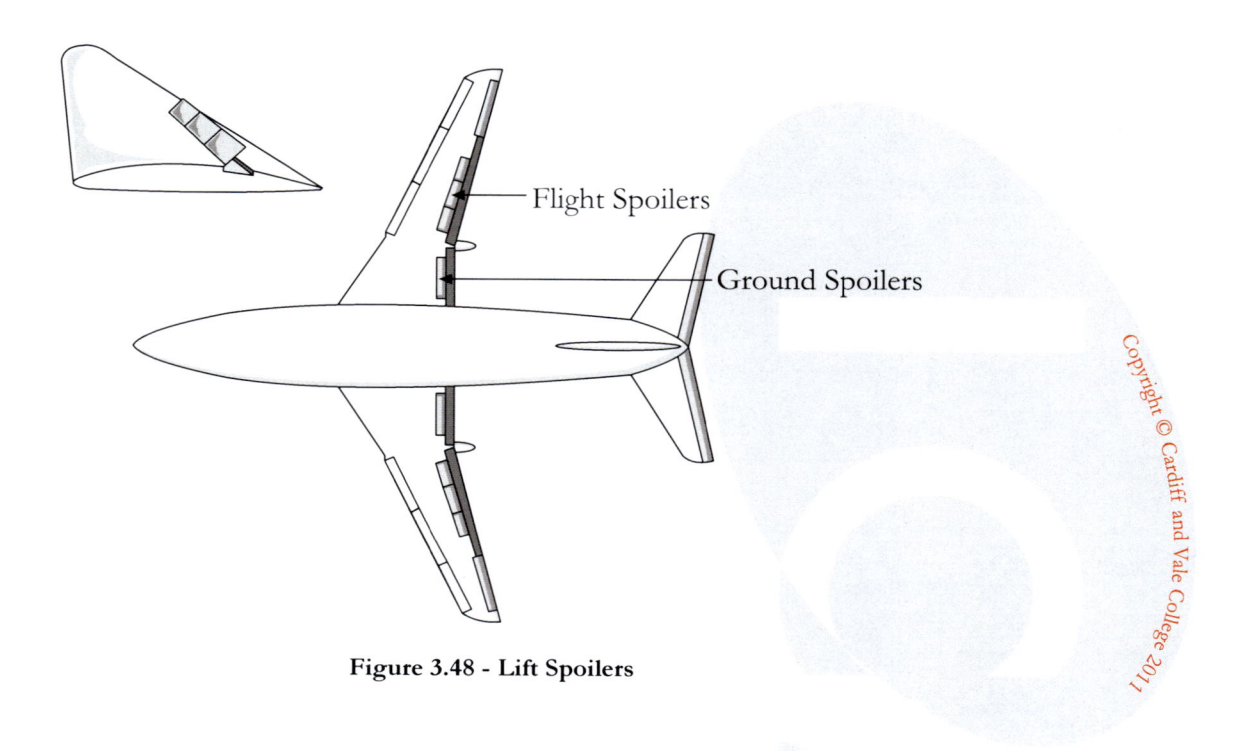

Figure 3.48 - Lift Spoilers

The flight spoilers have a dual use. The port and starboard flight spoilers can be deflected differentially in a limited flight range of travel to give roll control in support of the ailerons or, they can be deployed collectively to act as speed brakes in the air. If used in the latter role, differential deflection can still be used to maintain roll control during the air braking. Used as speed brakes, spoilers use the reduction in lift and the increase in drag to enable the aircraft to lose altitude rapidly without gaining airspeed. There are several speed brake selections each giving a different degree of spoiler deployment and thus a different degree of air braking.

When the aircraft lands, the ground and flight spoilers are both extended to full deployment to increase the drag on the aircraft and to destroy lift quickly so that the aircraft weight can be quickly transferred from the support of the wings onto the wheels.

Vertical Lift Control

If the flight spoilers are slightly extended, the aircraft can be trimmed to fly in this configuration. Now we have vertical height control. Retract the spoilers and the lift immediately increases and the aircraft rises. Deploy the spoilers further and the aircraft will sink.

Active Load Alleviation

This is a system that employs accelerometers to detect vertical acceleration movements of the aircraft in turbulence. If the lift of the wings continually fluctuates because of wind gusts, the structural loads fluctuate and the passengers get sick! By linking the accelerometer signals through a signal conditioner to issue commands to the spoilers the vertical accelerations are damped out. If a wind gust increases lift on the wings, the spoilers flick up to reduce it. The spoilers are fast acting and can move to deal with short cycle rapid lift changes. Use of load alleviation systems has had a favourable impact on the fatigue life of aircraft structures. Passengers like them as well!

Secondary Flight Controls

The lift augmentation controls are collectively referred to as the secondary flight controls. These include the wing flaps, leading edge devices, spoilers and the horizontal stabiliser. Incorporated with these controls are: The stall warning system, the unsafe takeoff warning system, the asymmetric flap control system and the flap load limiter system.

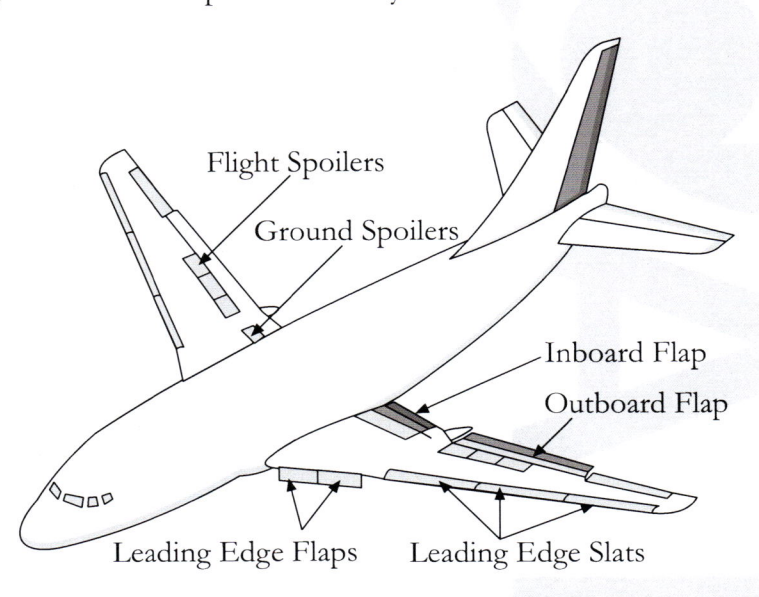

Figure 3.49 - Example of Secondary Flight Controls

CHAPTER THREE
THEORY OF FLIGHT

The modern commercial aircraft has an array of trailing edge devices along the whole of the span. You should now be able to identify all of these. One exception in the illustration below is the low and high-speed ailerons. I will briefly describe their use only because they are in the picture! They are primary flight controls used for lateral control on large aircraft. On takeoff the low speed ailerons are used because of the roll moment they can exert due to their position at the wing tips. After takeoff, the aircraft accelerates and the aerodynamic loads on the low speed ailerons would tend to twist the light structure of the wing tips. For that reason, they are locked out of use after takeoff when the flaps are retracted and the roll control is passed to the high-speed ailerons that are inboard on a stronger section of the wings.

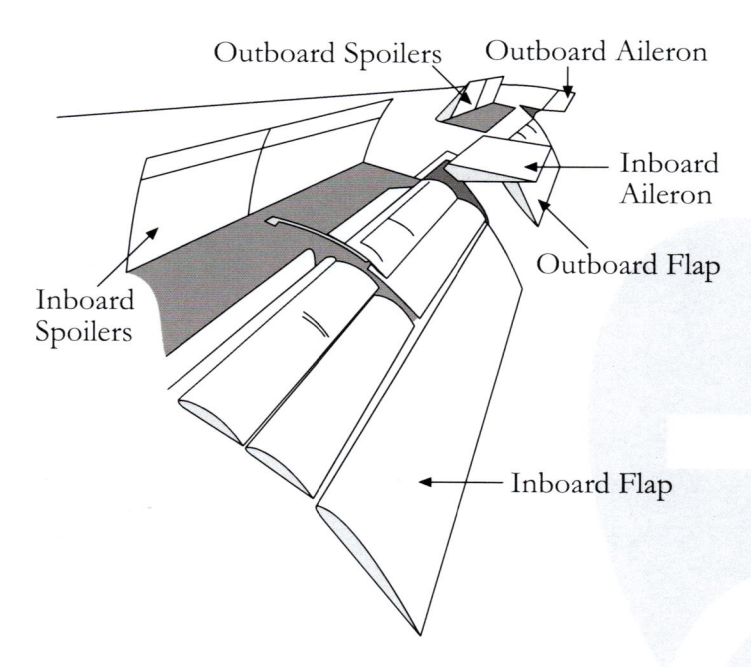

Figure 3.50 - Example of Trailing Edge Flight Controls

This is the end of the chapter. Do try the questions that follow this **and I strongly suggest a cup of coffee or what you fancy before embarking on the next and final chapter. See you there!**

Revision

Theory of Flight

Questions

1. Which of the following forces act on an aircraft in level flight:

    a. lift, thrust and drag

    b. lift, thrust and weight

    c. lift, thrust, weight and drag

2. If the nose of an aircraft is lowered in flight which of the following will occur:

    a. the angle of attack increases

    b. the rigger's angle of incidence decreases

    c. centre of pressure moves rearwards

3. The force opposing thrust is:

    a. lift

    b. drag

    c. weight

4. A glider travels forward under the influence of what force:

    a. thrust

    b. weight

    c. drag

5. If the lift-weight couple overcomes the thrust drag couple the direction of the correcting force on the tail plane is:

   a. upwards

   b. downwards

   c. equally split between up and down

6. If a single engine aircraft suffers an engine failure in level flight it will:

   a. pitch nose down

   b. pitch nose up

   c. not pitch either way

7. The arrangement of the lines of action of the forces on an aircraft in flight is:

   a. weight behind lift and thrust above drag

   b. thrust below drag and lift in front of weight

   c. thrust below drag and weight in front of lift

8. A Fowler flap will increase lift as a result of:

   a. increased camber and area

   b. increased angle of attack

   c. increased area only

9. When the trailing edge flaps are retracted in flight the attitude of the aircraft will tend to move:

   a. nose down

   b. nose up

   c. unaltered

10. When an aircraft is gliding at constant speed with no engine power:

   a. weight equals the resultant of lift and drag

   b. lift equals weight

   c. weight equals drag

11. When an aircraft is in a steady climb:

   a. lift equals weight

   b. thrust is less than drag

   c. lift is less than weight

12. To maintain height in level flight, if the engine power is reduced the angle of attack must be:

   a. increased

   b. decreased

   c. maintained

13. When an un-powered aircraft is gliding for maximum range the maximum:

   a. coefficient of lift is required

   b. rate of descent is required

   c. lift drag ratio is required

14. To achieve maximum range in a glide the airspeed selected should:

   a. be close to the stalling speed

   b. produce the maximum lift drag ratio

   c. produce the highest coefficient of lift

15. To achieve maximum endurance in a glide the airspeed selected should:

    a. produce the maximum lift drag ratio

    b. produce the least rate of descent

    c. give the steepest angle of glide

16. If drag is increased during a constant thrust descent it will:

    a. increase the rate of descent

    b. reduce the rate of descent

    c. not affect the rate of descent

17. When an aircraft is gliding for range its airspeed will be:

    a. lower than for endurance

    b. the same as that for endurance

    c. higher than for endurance

18. When an aircraft is in a level turn the:

    a. stall angle of attack is increased

    b. stalling speed is reduced

    c. stalling speed is increased

19. When an aircraft is in a level turn the engine power setting compared to that required for straight and level flight is:

    a. higher

    b. lower

    c. the same

20. An aircraft is conducting a level turn of 1000m radius at 100m/s the correct angle of bank for this turn is:

    a. 30°

    b. 45°

    c. 20°

21. The loading on a wing is calculated by:

    a. dividing the aircraft weight by the area of the wing

    b. dividing the area of the wing by the aircraft weight

    c. multiplying the weight of the aircraft by the wing span

22. If the proof load limit for an aircraft is exceeded the structure will:

    a. break

    b. be unaffected

    c. deform

23. A leading edge slat is designed to:

    a. generate lift only

    b. increase the stalling angle

    c. increase the stalling speed

24. The stalling speed of an aircraft during a long duration flight will:

    a. remain constant

    b. increase

    c. decrease

25. A leading edge flap is designed to:

    a. increase camber

    b. delay separation

    c. prevent stall

26. A Fowler flap:

    a. increases camber only

    b. increases wing area first then camber

    c. increases camber first and then area

27. A plain flap will increase:

    a. camber and area

    b. area only

    c. camber only

28. When an aircraft stalls it will:

    a. pitch nose up

    b. remain level and sink

    c. pitch nose down

29. When an aircraft is flying straight and level in still air the load factor (n) is:

    a. 1

    b. 0

    c. −1

30. When an aircraft is conducting a climbing turn the angle of attack increases on:

    a. the wing on the inside of the turn

    b. the wing on the outside of the turn

    c. both wings

31. An aircraft is held in a circular path in a correctly balanced level turn by:

    a. the horizontal component of lift

    b. the vertical component of weight

    c. the horizontal component of thrust

32. A stall-inducing strip would be fitted:

    a. span-wise on the leading edge of the wing tip

    b. chord-wise on the maximum camber of the wing root

    c. span-wise on the leading edge of the wing root

33. If an aircraft is flying into a headwind at a constant IAS its range will be:

    a. the same as in still air

    b. less than in still air

    c. greater than in still air

34. When the trailing edge flaps are extended in level flight the effective angle of attack will:

    a. increase but the stall angle will decrease

    b. decrease but the stall angle will increase

    c. exceed the basic stall angle

35. A low wing loading will:

   a. decrease the stalling speed

   b. increase the stalling speed

   c. not affect the stalling speed

36. When the trailing edge flaps are extended in flight the lift/drag ratio:

   a. increases

   b. is not affected

   c. decreases

37. When an aircraft is in a steady powered glide:

   a. lift equals weight and thrust exceeds drag

   b. thrust equals drag and lift equals weight

   c. weight exceeds lift and drag exceeds thrust

38. If the weight of an aircraft is increased the effect on its angle of glide and its gliding range are:

   a. angle of glide is increased and range is decreased

   b. neither is altered

   c. angle of glide is reduced and range is extended

39. When an aircraft enters a steady banked turn the angle of attack must be:

   a. increased

   b. decreased

   c. the same as for straight and level flight

40. During a steady banked turn the stalling angle:

    a. increases

    b. decreases

    c. does not alter

Revision

Theory of Flight

Answers

1. C
2. C
3. B
4. B
5. B
6. A
7. C
8. A
9. B
10. A
11. C
12. A
13. C
14. B
15. B
16. A
17. C
18. C
19. A
20. B

21. A
22. C
23. B
24. C
25. A
26. B
27. C
28. C
29. A
30. B
31. A
32. C
33. B
34. A
35. A
36. C
37. C
38. B
39. A
40. C

Flight Stability & Dynamics

Introduction

When an object is at rest, in other words motionless, and if it is in a state of equilibrium or balance, it will tend to remain at rest unless an external force acts on it to change that state. In the same vein, if the object is travelling in a straight line at a constant velocity it will tend to continue doing just that unless an external force acts on it to change either its direction and/or its velocity. This is the essence of Newton's first law of motion. The subject of stability is concerned with what occurs to an object once any external disturbing forces have been removed. In this sense we need to examine two things. First, the ability the object has to return to its original state after being disturbed. This is known as its *Static Stability*. Second, the ability the object has to return to its original state with the minimum of oscillation around it. For example, a pendulum that is disturbed from its original position of rest would continually overshoot the rest position in its attempts to regain it. Here the object will be governed by inertial and time-dependent affects. This is known as its *Dynamic Stability*. We will look at each of these stabilities in turn.

Static Stability

If you managed to balance a football on the tip of your index finger you could claim that for a moment at least, the ball was perfectly balanced. It would, however, be completely unstable. A simple shove would send it plummeting towards the floor! One cannot imagine that it would be able to recover its position on the tip of your finger once its balance had been disturbed. If it fell whilst suspended over a high cliff it would just keep accelerating on down to the beach. It just wants to keep on travelling away from you. Initially, it sat on your finger in a balanced state with no residual turning moments to make it do otherwise. Once you gave it a shove it moved and continued to move in a direction governed both by your push and the turning moment then exerted by its displaced centre of gravity position in relation to your finger.

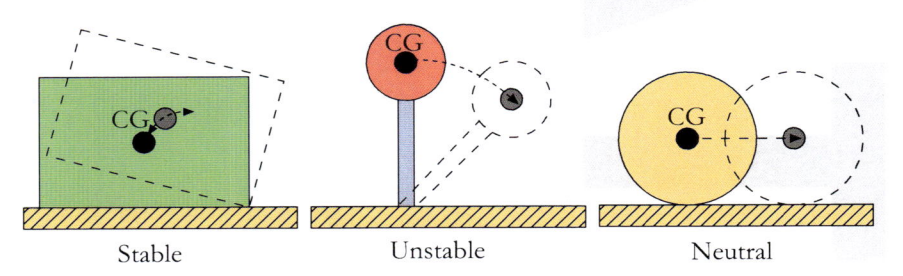

Figure 4.1 - Bodies in Various Forms of Equilibrium

CHAPTER FOUR
FLIGHT STABILITY & DYNAMICS

Things would be different if you tilted a wooden crate onto one edge and then released it. The crate would immediately return to the floor and its original rest position, provided you got your feet out of its way! The static stability in this case is of a very high order. There is another case that differs from both of the foregoing examples. Imagine the football at rest on level ground. You give it a gentle kick and it rolls a few yards and stops. The ball makes no attempt to return to its original position or to keep on moving away but takes up a new position where it is again balanced and at rest. The static stability in this case is neutral.

We can sum up these three classifications for static stability in regard to an aircraft in flight. A statically stable aircraft once disturbed from its original attitude of flight will naturally return itself to that original attitude. This may also be called positive stability. A statically unstable aircraft once disturbed from its attitude of flight will continue to diverge away from that position in the direction of the disturbing force. This may also be called negative stability. An aircraft that possesses neutral stability will, once disturbed, just take up the new position and remain there.

The terms positive stability and negative stability refer to stable and unstable respectively so do not think you are being confused if you come across the different descriptive terms in various texts on stability. We will use the latter descriptions as they are found in more common usage. Neutral stability remains just that, neutral.

Before we examine the static stability of an aircraft we need to look at the possible movements of an aircraft in terms of its axes of motion. There are three axes, each passing through the centre of gravity. These are, the longitudinal axis running from the fuselage nose to tail, the lateral axis running across the aircraft at right angles to the longitudinal axis and the normal axis that runs vertically through the centre of gravity.

Motion around these axes and the stabilities involved are shown in Table 4.1 below:

Axis	Motion Around Axis	Stability
Longitudinal	Roll	Lateral
Lateral	Pitch	Longitudinal
Normal	Yaw	Directional

Table 4.1 – Motion Around Axes

CHAPTER FOUR
FLIGHT STABILITY & DYNAMICS

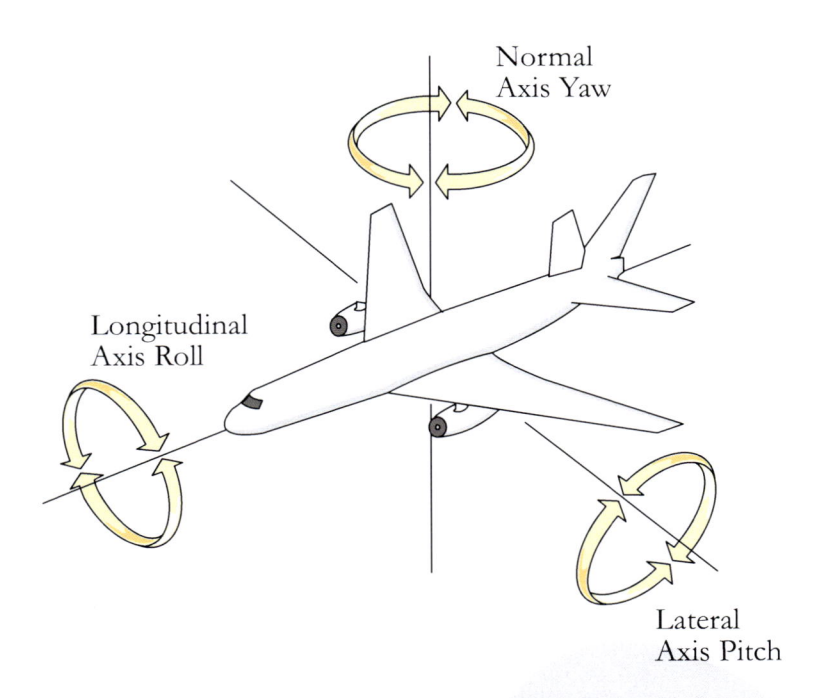

Figure 4.2 - The Three Axes of Movement

You need to learn these relationships because they are favoured questions. For example an aircraft displays lateral stability around the longitudinal axis. A disturbing force may induce a roll and the aircraft then rotates around its longitudinal axis. If the aircraft is stable it will return to wings level again. This stability is called lateral because it is the lateral axis that has actually been disturbed. Now examine the others and see if you can understand how they relate. Note that a disturbance around the normal axis creates a yaw. The stability is called directional because the aircraft heading is altered. It is not called longitudinal as you might have expected.

From what we have learnt so far we have established that balance is important but does not in itself produce stability. Imagine an aircraft flying straight and level with no residual turning moments around any axis. The aircraft is described as being 'trimmed' when in this situation. You would have no difficulty in believing that it would continue to fly straight and level forever, or as long as the fuel held out! But we can change that view if you now imagine that its centre of lift is acting right through its centre of gravity position. This is comparable to an aircraft on the ground that is precariously balanced on the point of a spike. It may appear to be in perfect balance but a sparrow landing anywhere other than on the centre of gravity position will unbalance the aircraft by producing a turning moment around one of the axes. The centre of gravity position will become misaligned with the centre of lift and produce a further turning moment that will cause the craft to fall off its pointed support.

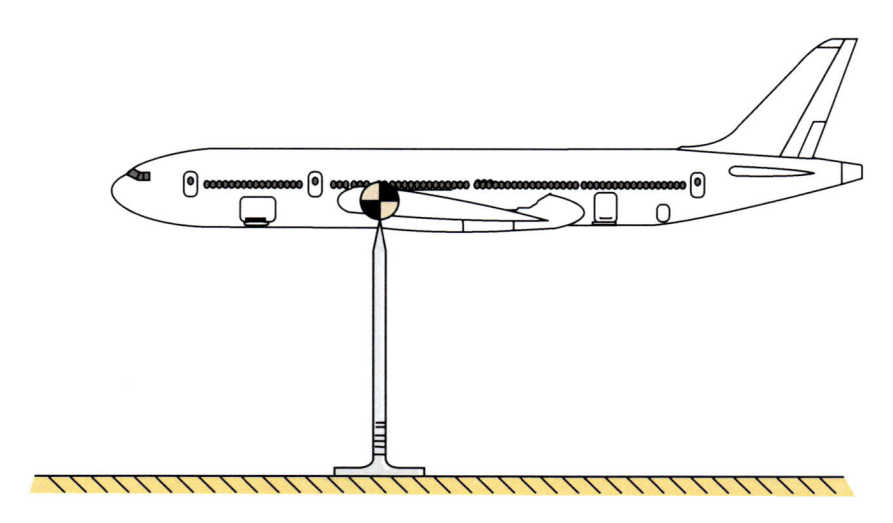

Figure 4.3 - Perfectly Balanced but Very Unstable

It is evident from the above that a designer needs to establish some means to stabilise the aircraft around each of its three axes of rotation. We will examine each of these stabilities in turn.

Longitudinal Stability

Longitudinal stability is created around the lateral axis. If an aircraft that is flying straight and level experiences an un-commanded nose up or down pitching motion then it is its longitudinal stability that will determine its ability to return to level flight again. You can visualise that the aircraft as viewed in side elevation is like a beam that is balanced on its centre of gravity. The lift acting through the centre of pressure in our illustrated case is exerting an anti-clockwise turning moment Lx1 around the lateral axis to pitch the nose down. This is being countered by the sum of two clockwise turning moments, $(T \times 2) + (F \times 3)$. The drag D in this case is aligned with the centre of gravity and will not impose a moment.

$$(L \times 1) = (T \times 2) + (F \times 3)$$

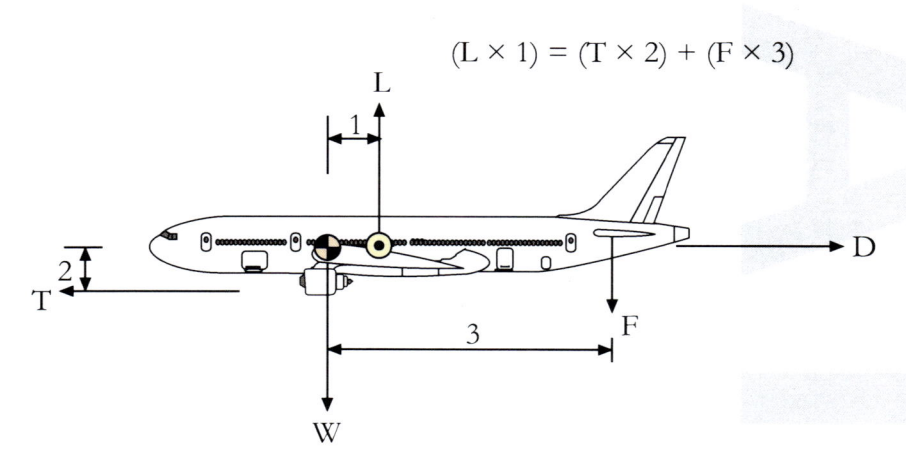

Figure 4.4 - Balance of Turning Moments

The illustrated case is one where static stability can exist. If the aircraft were to pitch down as a result of an increase in lift L × 1 the change in the tail plane angle of attack would produce more downward force on the tail plane to strengthen the correcting moment F × 3. Conversely if the nose pitched up the downward force on the tail plane would reduce or even reverse as it began to develop lift to produce an anti-clockwise moment to push the nose down again. So it seems that the angle of attack of the tail plane or horizontal stabiliser as our American friends like to call it is the governing factor in producing longitudinal static stability. You would be correct in assuming this but it does depend on a number of other conditions being favourable.

The lift or conversely the down force produced by the tail plane will also depend on its area and aerofoil shape. The turning moment it can exert will then depend on its longitudinal distance from the aircraft centre of gravity position. The bigger the distance, the less force it needs to exert to produce the required turning moment and the less area it will require to do this. This latter point is important because the tail plane also contributes to drag and should therefore not be unnecessarily big. The tail plane correcting force is the product of the tail plane area and the distance from its centre of lift to the aircraft's centre of gravity. Because area multiplied by distance gives units of volume this is referred to as the '*tail volume*'. The tail plane in normal trimmed flight is not required to produce much lift or down force. In the conventional case it produces a small down force to counter the turning moment exerted by the lift. This means that the total lift produced by the aircraft will be less than that produced by the wings alone. The tail plane down force has reduced it. Another reason to limit its influence.

The tail plane will need an angle of attack and be of sufficient area in order to produce the required forces. In normal trimmed level flight the angle of attack of the tail plane is conventionally set to a small negative angle of incidence to produce a down force. Sounds good but we need to be aware that the relative positions of the aircraft's centre of gravity and centre of pressure have a direct affect on longitudinal static stability. If you look at a case where the centre of gravity were to be positioned behind the centre of pressure for example then the turning moment produced by lift would tend to pitch the nose up. This would increase the angle of attack of the wings and as you have read earlier, the centre of pressure would then move further forward and will in this case strengthen the pitch up moment in a de-stabilising effect. The increase in downwash angle produced by the wings will then affect the tail plane and reduce its effective angle of attack thus reducing its lift. If the tail plane cannot produce sufficient force to correct this situation we will have a big problem. In any event the situation will be one of static instability.

CHAPTER FOUR
FLIGHT STABILITY & DYNAMICS

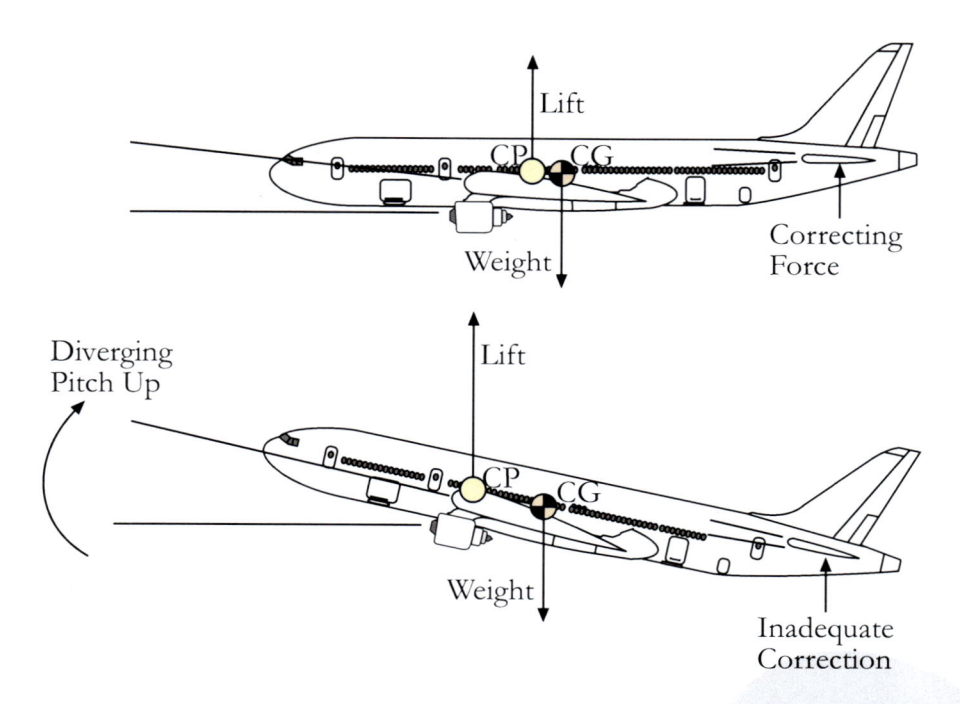

Figure 4.5 - Unstable Effect When Centre of Gravity is Too Far Back

On a conventional aircraft the centre of pressure will be positioned behind the centre of gravity. You have already read that stability reduces the closer the centre of pressure comes to the centre of gravity. The aircraft becomes more responsive to control movements as a result of this so it is not entirely undesirable providing it does not get too close. You would not want an aircraft that was so stable that it refused or made it difficult for you to change its attitude when you required it to! A further advantage to the conventional positioning is that when the aircraft pitches up, the centre of pressure will move forwards but the increase in lift will tend to strengthen the pitch down moment. This is a stabilising effect. The tail plane is thus assisted in its task of bringing the nose down again. Conversely, if the nose were to pitch down, the centre of pressure will move rearwards but the reduced lift will tend to weaken the pitch down moment. Again this is a stabilising effect.

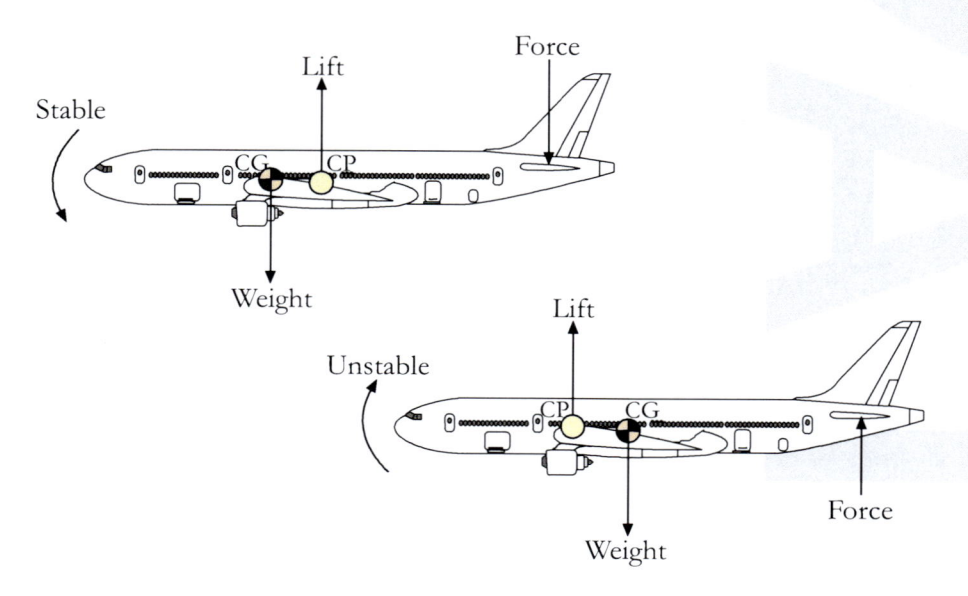

Figure 4.6 - Longitudinal Stability

If we now return to the image of an aircraft that has coincident centre of pressure and centre of gravity positions we can begin to understand the stability problems this would incur. As the nose pitches up, the centre of pressure moves ahead of the centre of gravity increasing the pitch up moment. As the nose pitches down the centre of pressure now moves behind the centre of gravity increasing the pitch down moment. The forces exerted by the tail plane will have to alternate in an attempt to correct this. Somewhere in the middle of this we would be holding our breath! Static stability will be poor and the aircraft will be very 'skittish' or over-responsive to any pitch control movements. Not much fun if you are trying to attempt a landing! Imagine your car having a step up ratio in its steering gear, a twitch of the steering wheel and you turn half a circle! Military aircraft actually benefit from this by being more agile but those aircraft types cannot be controlled manually as the human brain cannot respond quickly enough to the oscillatory movements experienced by the aircraft in its attempts to stabilise in new attitudes.

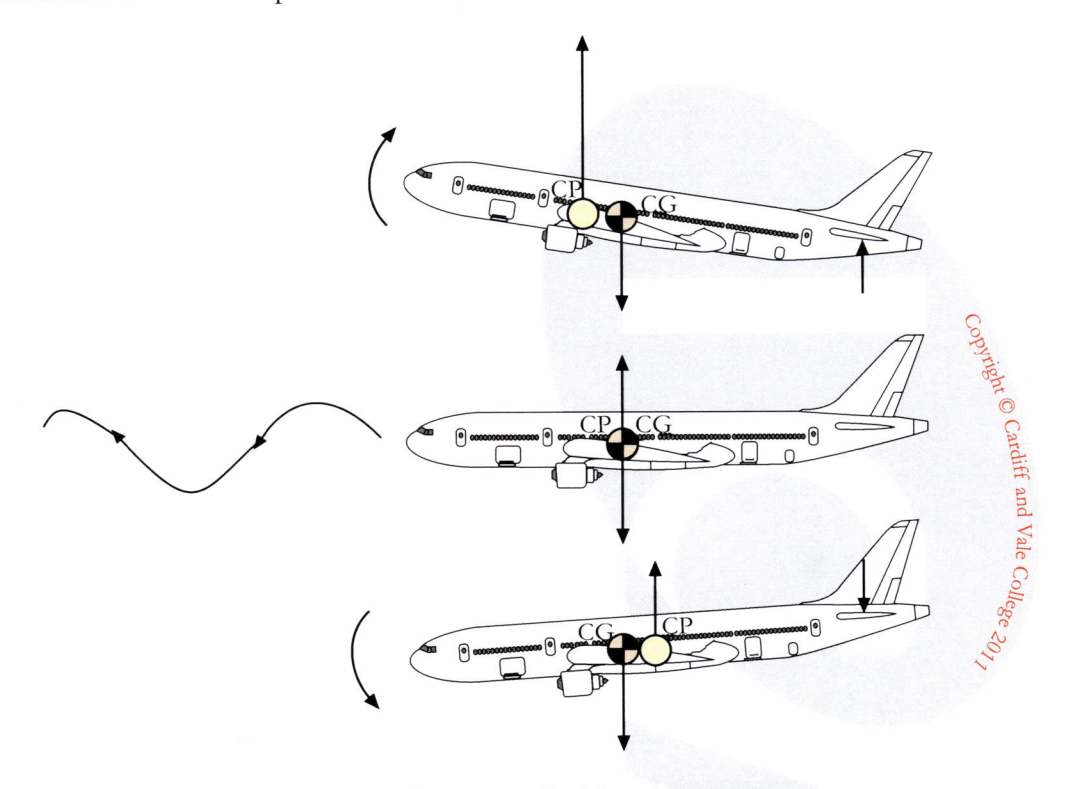

Figure 4.7 - Effect of Coincident C of G and CP Positions

Longitudinal Control

As we have encountered the topic of control in pitch it is worth looking at the control surface used to achieve it. The elevator fitted at the rear of the tail plane effectively changes the camber of the tail plane and thus its lift. If the elevator is moved up by pulling back on the flight deck control column the negative camber created will exert a down force and pitch the aircraft up.

Conversely, pushing the control column forward to achieve a down movement of the elevator creates positive camber and pitches the aircraft down. Some configurations of tail plane have a variable incidence in that the tail plane itself can be moved to exert more or less force. This feature is used for pitch trimming purposes, not for control. If the tail plane is selected to negative incidence or leading edge down for example this will create a down force to create a pitch up triming moment on the aircraft.

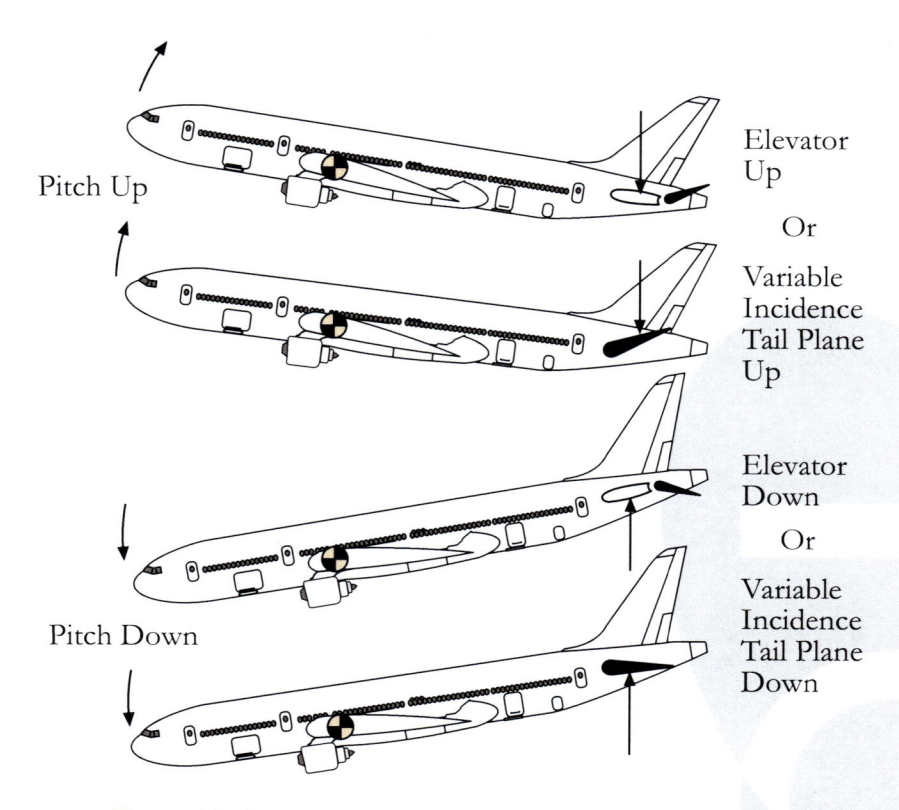

Figure 4.8 - Longitudinal Movement around the Lateral Axis

Movement of the elevators will create a turning moment to override the restoring moment that is the natural stability of the aircraft. As the longitudinal stability is affected by the position of the aircraft's centre of gravity you may appreciate that the further back this point is, the closer it approaches to the centre of pressure and the easier it becomes for the elevator to overcome the restoring moment to command a change in pitch attitude. Conversely, the further forward the centre of gravity is, the further it will be from the centre of pressure position and the harder it becomes for the elevator to override the aircrafts natural stability. Note that the further forward the centre of gravity is, the more longitudinally stable the aircraft becomes and the further back it is, the less stable it is.

Neutral Point & the Static Margin

It should be now apparent to you that there has to be some control exerted over the position of the centre of gravity in flight. There is a range published in the Operating Manual that states the permitted maximum limits for the forward and rear positions of the centre of gravity. The forward position will be governed by balance and the degree of control response required. The rearward position is set to avoid reaching the *neutral point*. This is the point where the centre of gravity position is so far aft that the increased tail down moment produced by this can only just be matched by the tail plane restoring moment. The tail plane moment will have reduced because of the decreased moment arm. When this situation is reached the longitudinal stability becomes neutral. If the nose drops the aircraft will remain in the pitched down attitude. Any further movement of the centre of gravity to the rear will result in instability. The distance between the centre of gravity position at any time and the neutral point is called the *static margin* and is a measure of the longitudinal static stability of the aircraft.

Manoeuvre Point & the Manoeuvre Margin

The further back the centre of gravity is positioned the less longitudinally stable the aircraft becomes. This means the aircraft tends to become over responsive to control movement. A point is reached in the rearward positioning of the centre of gravity where the control response becomes unacceptable. This is called the *manoeuvre point* and the distance between the centre of gravity position at any time and this point is called the *manoeuvre margin*. The manoeuvre point will be close to the neutral point in most cases.

Trim Drag

When the centre of gravity is well forward the tail plane has to produce an increased down force. In doing this it also experiences increased drag. This is called trim drag. As the centre of gravity is moved rearwards the force required to stabilise the aircraft reduces and thus the trim drag reduces. The area of the tail plane will also influence trim drag.

Longitudinal Dihedral

For a conventional aircraft to attain a position of longitudinal trim in the normal flight range it is necessary to set the tail plane angle of incidence to a smaller angle than that of the wings. The difference between the incidence angles of the wing and tail plane is known as the *longitudinal dihedral*. In the case where the tail plane angle of incidence is less than that of the wing it is referred to as positive dihedral. The further forward the centre of gravity is the more the stabilising effect is and the greater will be the down force required on the tail plane. This means the tail plane must have a reduced angle of incidence or it will create too much lift.

For less conventional aircraft where the centre of gravity is positioned behind the centre of pressure the tail plane angle of incidence has to be set to give an increased up force or lift. This requires that the tail plane angle of incidence should be set to a higher angle than that of the wings. The difference between the tail plane and the wing angles of incidence in this case is known as negative longitudinal dihedral.

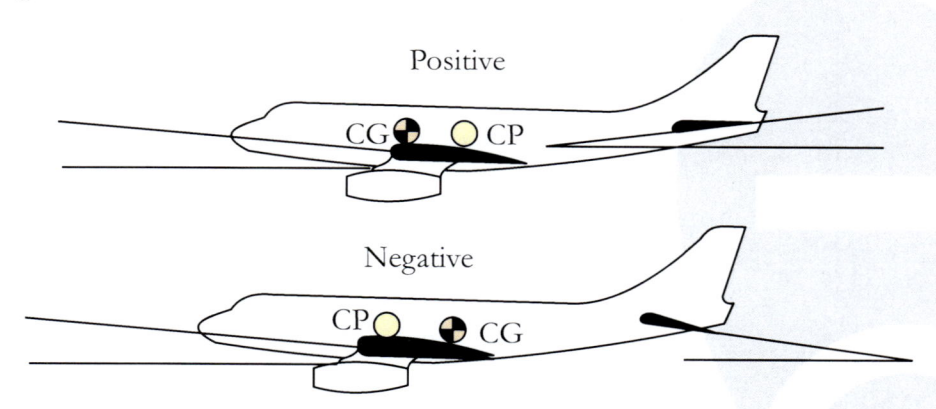

Figure 4.9 - Longitudinal Dihedral

Stick-fixed Stability

Up to now we have been assuming that the elevators remain in a fixed, centred position and cannot influence the stability of the aircraft. This known as '*stick-fixed*' stability and assumes the controls are being held centralised. If the aircraft is flown '*stick-free*' then the elevator will tend to trail or droop and this will reduce the down force on the tail plane upsetting the longitudinal trim. The main difference for us to know is that longitudinal stability is less in the 'stick-free' configuration than it would be if it were 'stick-fixed'.

Canard Configuration

We cannot end a discussion on longitudinal stability without examining the canard. In this configuration the tail plane is re-positioned to be ahead of the wings. When this is done it becomes a fore plane or, as it is correctly known, a canard.

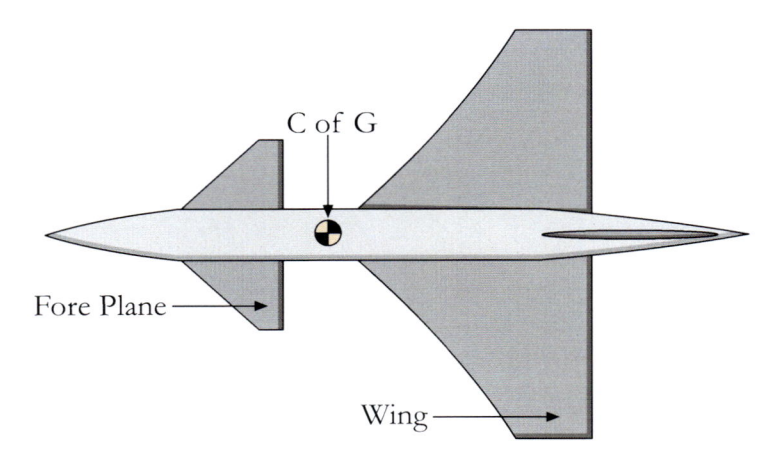

Figure 4.10 - Canard Configuration

The big difference apart from its position is that the fore plane now has to produce an up force or lift to create a stabilising restoring moment. This force supplements the lift produced by the wings thus increasing the total lift of the aircraft. The arrangement has positive longitudinal dihedral in that the fore plane needs to create lift so its angle of incidence unlike the tail plane on conventional aircraft is greater than that of the wing but is still referred to as being positive in this case. This has another advantage in that at the stalling speed it is the fore plane that stalls before the wings. This drops the aircraft nose preventing a wing stall. You could say a canard aircraft is pretty well stall proof! At least, in theory it is. Should both the fore plane and the wings stall it would be unlikely that the aircraft could be recovered.

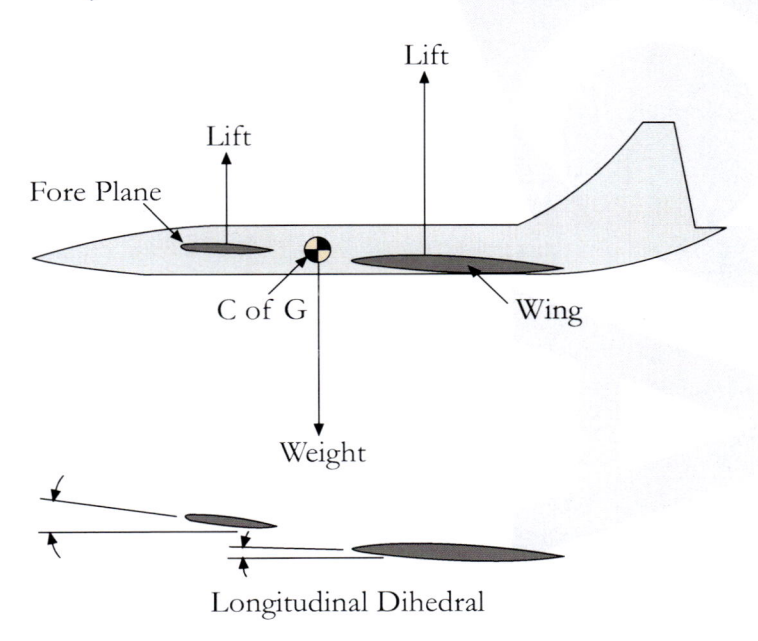

Figure 4.11 - Longitudinal Stability of a Canard Arrangement

Longitudinal control on a canard is achieved by changing the lift of the fore plane by varying its incidence.

High Speed Aircraft

There are a few considerations accompanying high-speed aircraft. As an aircraft passes through the transonic to the supersonic range there is a pronounced rearward shift of the centre of pressure. This creates a pitch down turning moment that has to be corrected. One way of achieving this is to automatically re-position the tail plane incidence to increase its down force and give a pitch up restoring moment. This is called '*Mach trimming*' and it creates additional trim drag. Another method that was used on the Concorde was the rapid transfer of fuel into a fin tank to move the aircraft centre of gravity position rearwards and thus restore the longitudinal trim. This avoided the creation of trim drag.

The swept back configuration also used on the tail plane reduces its coefficient of lift in just the same way as it affects the wings. This will reduce its effective restoring moment unless the tail plane incidence or area is increased. This increases the trim drag.

As you have seen, there is quite a lot to longitudinal static stability but I hope you did not have to struggle too much. A coffee break may be indicated before we launch into the topic of lateral stability.

Lateral Stability

Lateral stability is created around the longitudinal axis. If an aircraft is disturbed around this axis it will roll. This results in one wing descending whilst the opposite wing rises.

The stability of the aircraft should create a counter turning moment that will resist the roll and then produce an opposing roll motion to restore the aircraft to wings level again.

Roll Damping

To grasp what is happening during a roll motion I would like you to visualise the wing travelling forwards in the direction of flight while moving vertically either up or down at the same time. We can now examine these two movements and see what effect they have on the relative airflow direction. In the case of the down-going wing the change in relative airflow direction will increase the wing angle of attack thus creating additional lift. In the case of the up-going wing the angle of attack will reduce thus reducing lift. The combined effect of the changes in angle of attack and lift will be to create an imbalance in wing lift forces that acts to damp or resist the rolling motion. Note that this damping action will only exist whilst the aircraft is actually rolling and it cannot by itself restore the aircraft to wings level again. Once the roll has stopped the aircraft will be left in a banked position. In this sense roll damping is an illustration of neutral lateral stability. However, the aircraft will then sideslip under the influence of a side component of the inclined lift force.

The strength of the roll damping is proportional to the rate at which the aircraft rolls. A problem will occur if the aircraft is near the stalling speed when the roll occurs. The increase in the angle of attack on the lower wing may cause that wing to stall and the damping action will cease. The aircraft is then unstable and at risk of entering a spin.

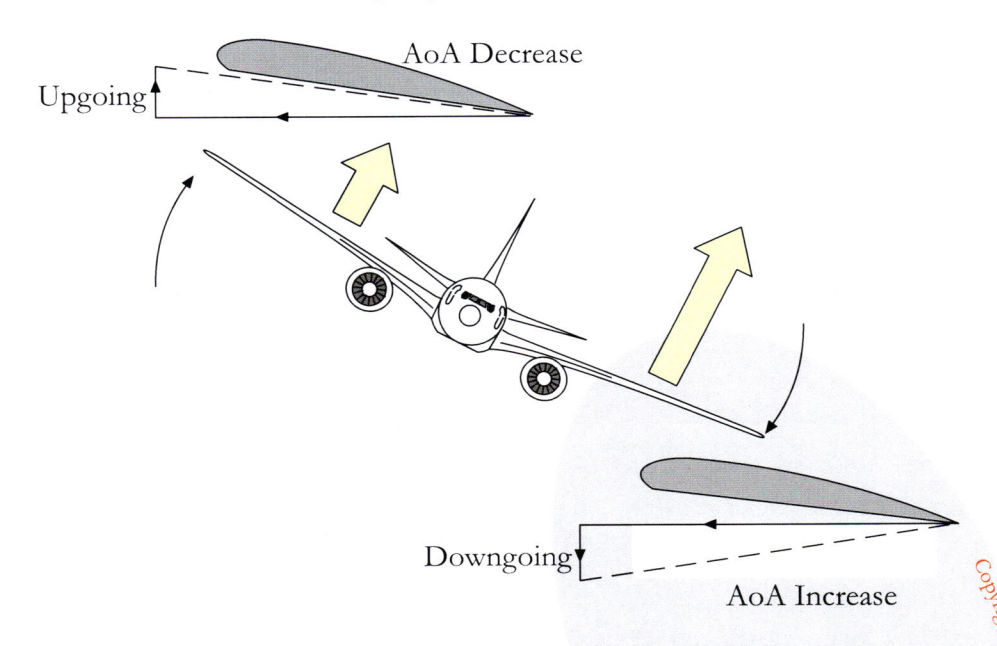

Figure 4.12 - Change in Effective Angle of Attack (AoA) During a Roll

Lateral Dihedral

The wings on a conventional aircraft are set so that they are inclined up from the horizontal along the span from root to tip. The angle between the span of the wing and the horizontal is called the *lateral dihedral*.

Figure 4.13 - Lateral Dihedral

Stability During a Sideslip

Imagine now that the aircraft is travelling forwards whilst slipping to the side and down all at the same time. The effect of the dihedral in this situation is that the relative airflow direction will produce a bigger angle of attack and thus increased lift on the lower wing and a reduced angle of attack and thus lift on the upper wing. The lateral imbalance created by the difference in lift between the upper and lower wings and the fact that the upper wing is now partly shielded by the fuselage from the airflow will roll the aircraft back towards the wings level position. Once the aircraft is restored to wings level the lift on both wings will be again equal.

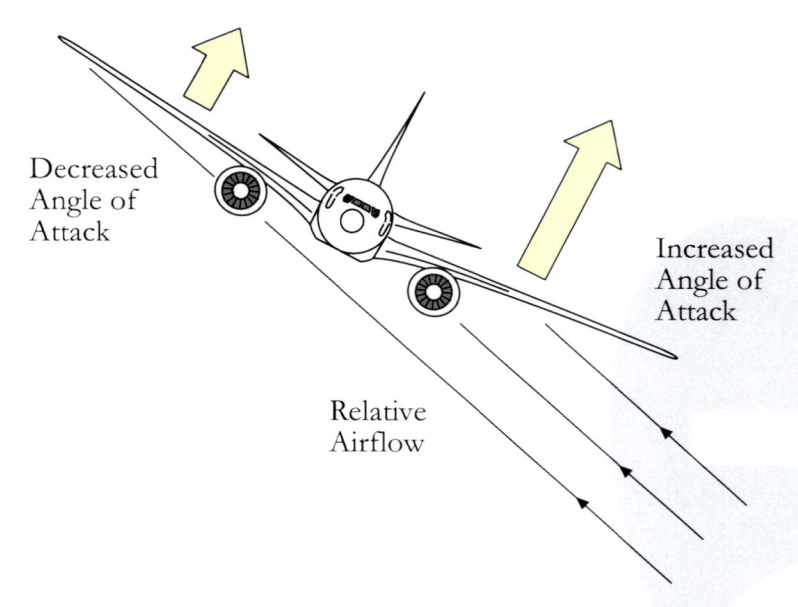

Figure 4.14 - Effect of Dihedral on Lateral Stability During a Sideslip

The dihedral effect gives an aircraft its lateral stability when it sideslips as a result of a roll or a yaw. When an aircraft yaws the wing on the outside of the turn moves forward whilst the inner wing retreats, this produces an imbalance in lift that causes the aircraft to roll and then sideslip. The difference between the aircraft's directional and lateral stabilities will influence the degree to which this will occur. High directional stability and low lateral stability will increase the tendency to roll during a yaw.

Effect of Sweepback on Lateral Stability

When an aircraft with rearward swept wings sideslips the lower wing is presented more normal to the airflow than the upper wing. This has the effect of producing more lift on the lower wing than that produced by the upper wing. The lower wing then acts as if it possessed a higher aspect ratio than the upper wing. The imbalanced lift forces resulting from this when added to the dihedral effect produce a very strong lateral restoring moment that rolls the aircraft back to wings level again. You can confidently say that sweepback improves lateral stability.

A problem with wing sweepback is that if the aircraft yaws the forward moving wing presents itself more normal to the airflow and thus increases lift. This induces a strong rolling movement and sideslip. A yaw will always produce a sharp roll on a swept wing aircraft.

A further problem encountered on swept wing aircraft is that the roll motion can result in a tip stall on the down-going wing. This is more likely to occur during low speed operations.

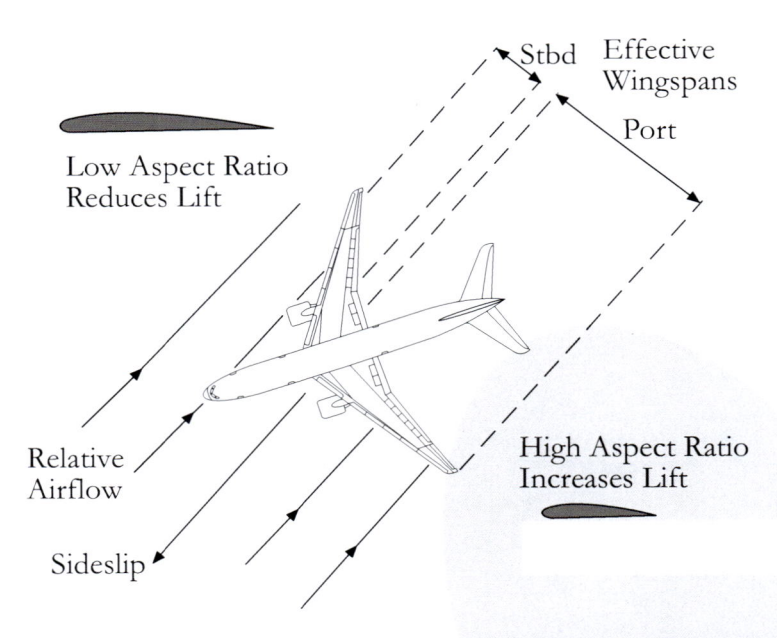

Figure 4.15 - Effect of Sweepback on Lateral Stability During a Sideslip

Anhedral

There is a tendency for swept wing aircraft to become over-stable laterally because of the combined effects of dihedral and sweepback. This is particularly evident at low speeds and high angles of attack. Every ten **degrees of wing** sweep is the equivalent of an additional degree in dihedral. Where over-stability becomes a problem, the aircraft designer may **reduce** the dihedral or even make it negative to produce anhedral. This reduces the **lateral static stability** of the aircraft and makes the aircraft more responsive to **lateral control.**

Figure 4.16 - Lateral Anhedral

Low Wing Aircraft

When a low wing aircraft is in a sideslip the airflow tends to create a downwash over the lower or leading wing and an upwash on the upper or trailing wing. This alters the effective angles of attack of each wing. The lower or leading wing effective angle of attack decreases and the upper or trailing wing angle increases. This produces a de-stabilising effect that tends to reduce the dihedral effect. Be clear though, the dihedral effect although decreased will be designed to overcome this and produce an overall stabilising influence. If the wings are mounted centrally on the fuselage sides the former effect is cancelled out and the dihedral effect is thus unaltered. The aircraft keel surface with the vertical fin in particular lying above the aircraft centre of gravity position on a low wing aircraft produces a significant lateral restoring moment. A large high fin exerts a strong restoring influence. The keel surface is described below.

High Wing Aircraft

There are a number of factors that affect high wing aircraft that give them an overall improvement in lateral stability. During a sideslip the fuselage causes the air to up-wash under the lower wing thus increasing its effective angle of attack and thus lift. The airflow under the upper wing tends to down-wash reducing the angle of attack and thus lift. This effect is equivalent to an increase in dihedral effect and the lateral stability is improved.

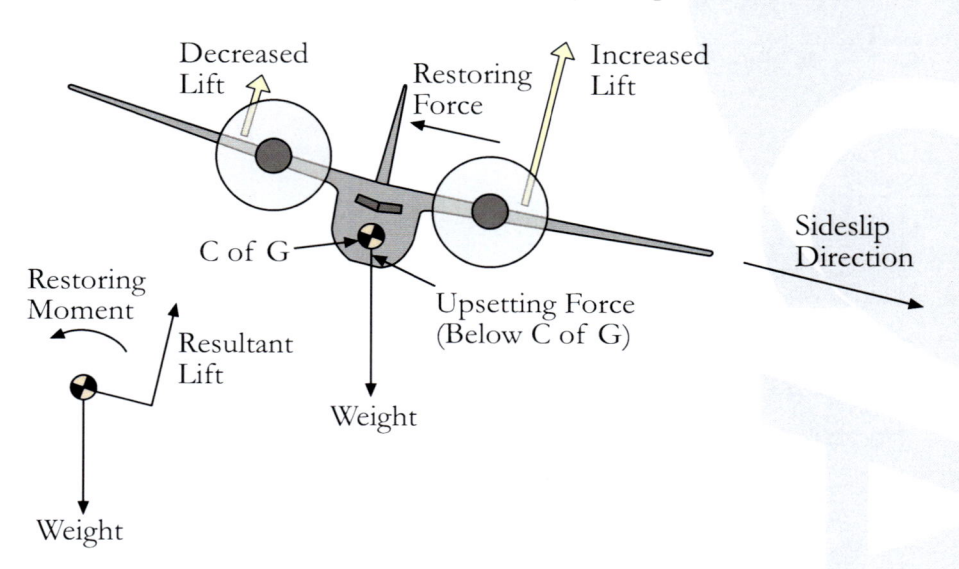

Figure 4.17 - Lateral Stability of a High Wing Aircraft

During the sideslip, the line of action of the total lift force moves across the lateral axis towards the lower wing. This puts the total lift line of action out of alignment with the line of action of weight. This creates a turning moment like a pendulum effect as the weight tries to align itself under the lift again and lateral stability is once again improved.

The component of the airflow that passes span-wise over the top surface of the wings during a sideslip creates drag that in turn creates a restoring moment. The effect of the aircraft's keel surface is also significant in improving the lateral stability of high wing aircraft. The keel surface is described below.

Keel Surface

The keel surface of an aircraft is the side surface area of the fuselage and the surface area of the fin that will be exposed to the oncoming airflow in a yaw or a sideslip. The surface area of the fin and the fuselage keel area situated above the centre of gravity will create a lateral correcting moment in a sideslip. The fuselage keel surface below the centre of gravity will create an upsetting moment. A large high fin and a low centre of gravity position will therefore improve the lateral stability. On a high wing aircraft this may improve lateral stability so much that the wings may require anhedral.

Components of the keel surface producing positive lateral stabilising effects are a high fin and rudder and in particular a 'Tee-tail' configuration where the tail plane is on top of the fin, a high wing and a low centre of gravity position.

The keel surface aft of the centre of gravity position also provides directional stability in restoring the aircraft heading following a yaw. Conversely, the keel surface ahead of the centre of gravity position would act to de-stabilise the aircraft during a yaw. We will re-visit this in the section on directional stability.

Shielding Effect of Fuselage

During a sideslip most aircraft are affected by the fact that the fuselage shields the inner part of the upper or trailing wing. The lift in the shielded area is reduced thus creating an imbalance in the lift of the wings that exerts a correcting moment that improves lateral stability.

Effect of Flap Deployment on Lateral Stability

If the flaps on the aircraft are situated on the inboard wing trailing edges the lift distribution of the wings will change significantly when the flaps are deployed.

The line of action of the resultant of the lift on each wing will move inboard on deployment of the inboard flaps. This means that the lateral turning moment exerted by the lift around the longitudinal axis will be reduced. The effect will be that the lateral stability of the aircraft will be reduced on deployment of the inboard flaps.

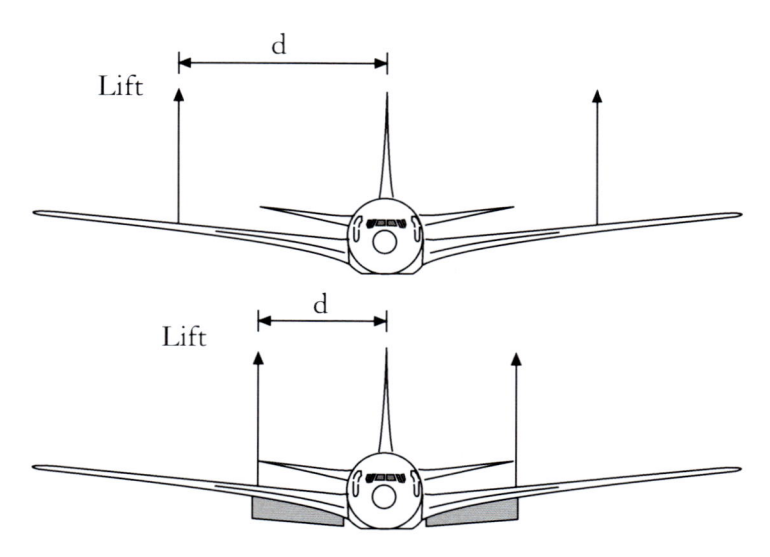

Figure 4.18 - De-Stabilising Effect of Flaps

Lateral Control

Lateral control around the longitudinal axis is carried out by use of the port and starboard ailerons. These are positioned on the trailing edges near to the wing tips so that they can exert maximum leverage around the longitudinal axis. When the control yoke on the flight deck is rotated to the left for example, the aircraft will roll to the left and vice versa. During a roll to the left, the port aileron will be deflected up spoiling the lift of the port wing in its area of influence and the starboard wing aileron will be deflected down increasing the lift of the starboard wing in its area of influence. The lift imbalance between the port and starboard wings will cause the aircraft to roll. The flight spoilers are often used to assist the ailerons and will partially deploy in support of the up-going aileron when large roll commands are made.

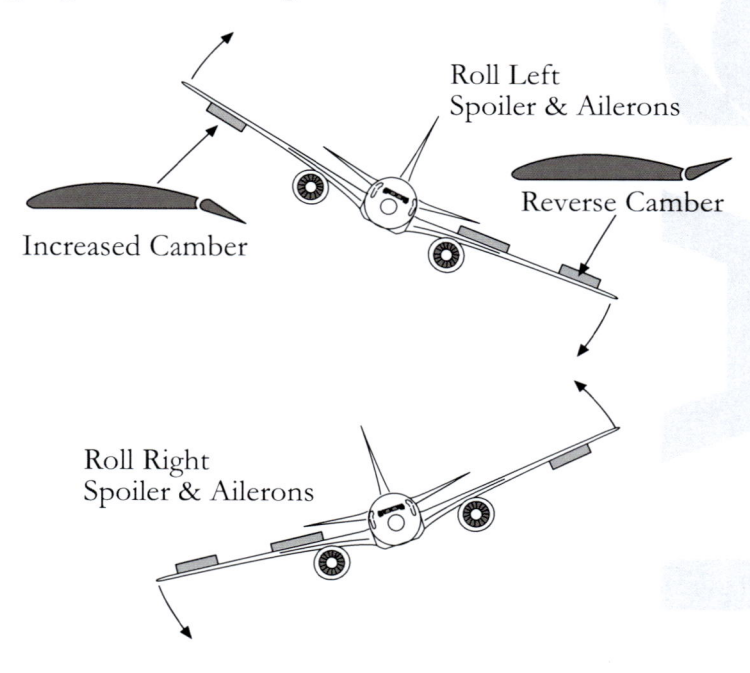

Figure 4.19 - Lateral Control Around the Longitudinal Axis

Adverse Yaw

A problem occurs when the ailerons are used to command a roll. The down-going aileron increases lift that in turn increases the induced drag of its parent wing. The up-going aileron decreases lift and that reduces the induced drag of its parent wing. The drag imbalance created causes the aircraft to yaw in a direction opposite to the commanded roll direction. For example, when moving the control yoke to the left to command a roll to the left the aircraft yaws to the right. This is without doubt unacceptable so the aircraft design incorporates measures to counteract adverse yaw.

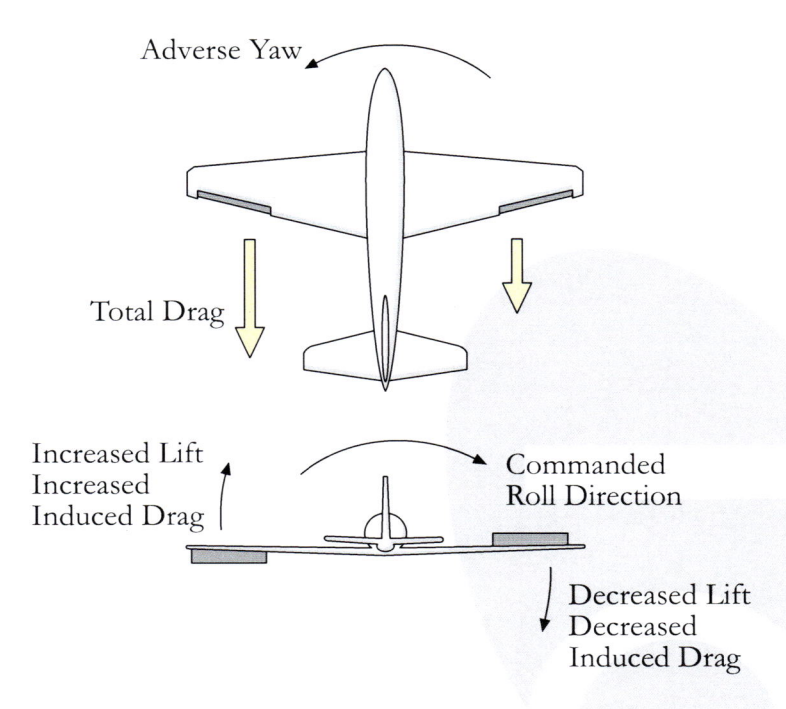

Figure 4.20 - Adverse Aileron Yaw

When the ailerons are used the drag imbalance is created by the induced drag only. The profile drag would be similar for each aileron if they each deflected by the same amount. In order that adverse yaw may be avoided an opposing imbalance is created in the profile drag produced by each aileron. This is done in one of two ways.

Frise ailerons are specially contoured to project a 'beak' into the airflow underneath the wing of the up-going aileron. This creates additional profile drag to compensate for the reduction in induced drag on the wing with the up-going aileron.

Differential ailerons on the other hand are rigged so that the up-going aileron deflects up a greater number of degrees than the down-going aileron deflection. This device uses profile drag to balance out the difference in induced drag and thus prevent adverse yaw.

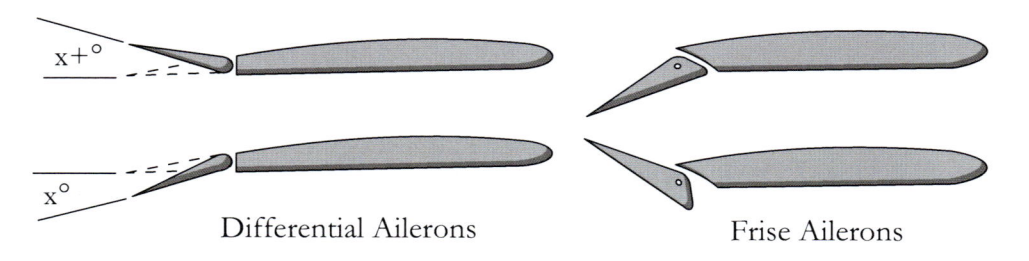

Figure 4.21 - Methods used to Overcome Adverse Yaw

Directional Stability

This is the aircraft's static stability around its normal axis. A disturbance around the normal axis will cause the aircraft to yaw either to the left or to the right. If you have ever thrown a dart you will have witnessed its directional stability but have you considered what produces it?

Weathercock Effect

The feathered vanes at the rear of a dart act in the same manner as a weathercock. The centre of gravity position is well forward of the vanes giving them the ability to exert a powerful turning moment around the normal axis. The normal axis passes through the centre of gravity. If the dart alters heading in a yaw around the normal axis the vanes will be presented at an angle of attack to the relative airflow and an aerodynamic force is created. The force acts on the moment arm to the centre of gravity position and the dart rotates back around its centre of gravity until the original heading is restored.

Like the dart an aircraft has a fin or vertical stabiliser that acts like the feathered vanes. The area of the fin and its distance from the centre of gravity position determines the directional stability. The fin has a symmetrical aerofoil section.

Effect of Keel Surface

The keel surface to the rear of the centre of gravity position contributes to the weathercock effect. A large keel surface including a large area fin will have a powerful stabilising effect. However, the keel surface ahead of the centre of gravity position will exert a de-stabilising effect. It is important that this surface is not so predominant that it counteracts the weathercock effect.

Relative Airflow

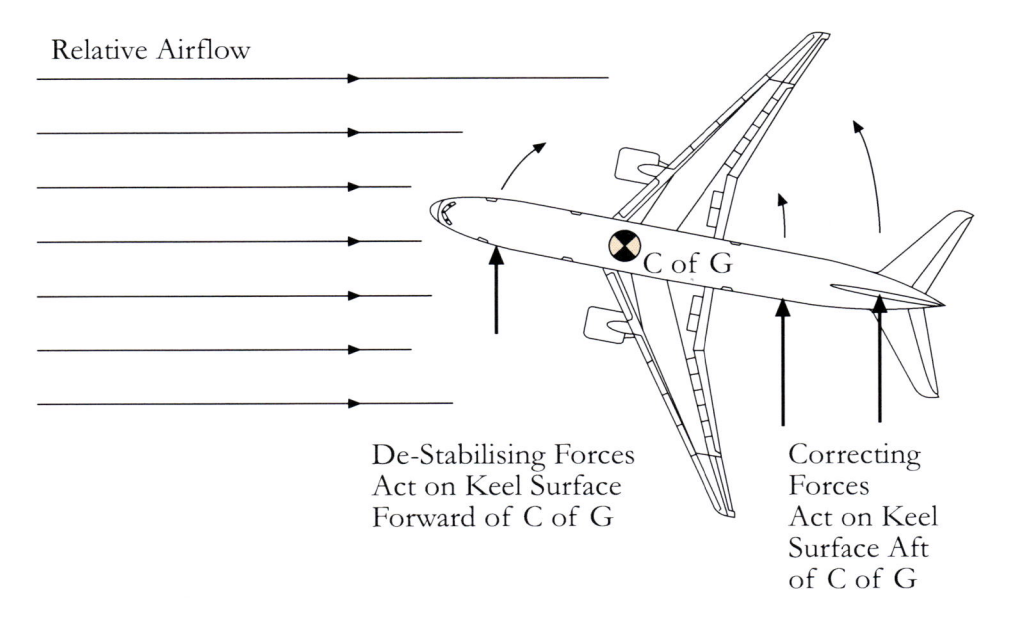

De-Stabilising Forces Act on Keel Surface Forward of C of G

Correcting Forces Act on Keel Surface Aft of C of G

Figure 4.22 - Directional Stability

Longitudinal Centre of Pressure

When an aircraft yaws or sideslips the airflow striking the keel surface can be said to create forces that have a resultant that acts through a given point on the keel. It is important that the longitudinal centre of pressure position is to the rear of the centre of gravity position. The area of the fin will have been chosen to ensure that this is the case.

Design of the Fin

The fin or vertical stabiliser has a symmetrical aerofoil shape that will produce an aerodynamic force when it is presented at an angle of attack to the relative airflow on either side. At high angles of attack such as those encountered during a steep sideslip the fin can stall. This reduces the side force and de-stabilises the aircraft in yaw. To increase the stalling angle of the fin it is usually designed to be swept back and have a low aspect ratio.

Tee-tail aircraft have the horizontal stabiliser mounted on top of the vertical stabiliser. This configuration makes the fin more effective because its tip is capped.

Centre of Gravity Position

The further forward the aircraft centre of gravity position is, the more the directional stability of the aircraft is increased. Conversely the more aft the centre of gravity position is, the less the directional stability.

Shielding Effect

At very high angles of attack the fuselage can shield the fin from the relative airflow. Because of the very high angles of attack adopted by swept wing and slender delta aircraft configurations at low air speeds they are prone to directional instability.

High Speed Aircraft

At supersonic speeds the aerofoil surface is less effective than at low speeds. For this reason supersonic aircraft require proportionately larger fin areas.

Directional Control

Directional control around the normal axis is carried out by **use of the rudder**. When the pilot pushes the left rudder pedal forwards the **rudder deflects to the left** causing the aircraft to yaw left and vice versa.

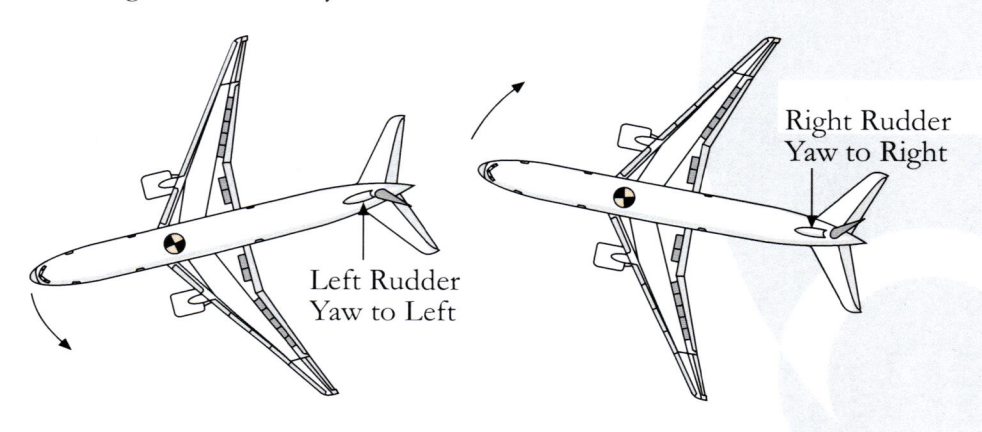

Figure 4.23 - Directional Control Around the Normal Axis

As the aircraft speed increases the range of movement of **the rudder** is automatically reduced to just a few degrees. This is done to **prevent the** aerodynamic force on the rudder becoming too large and causing structural damage.

Cross Links Between the Axes of Rotation

We have been examining each of the stabilities in turn. There is however a strong connection between the roll stability and the yaw **stability and on** occasions these can combine to form a weaker link with the longitudinal stability. The overall effect is called spiral stability.

When an aircraft experiences a yaw the lift imbalance created between the forward going wing and the rearward going wing causes the aircraft to roll in the direction of the retreating wing. The aircraft then sideslips and this creates a side force on the fin and keel surface that induces a further yaw causing the aircraft nose to drop. A yaw will create a roll.

Swept wing aircraft are particularly prone to this reaction. When a swept wing aircraft yaws the forward moving wing is presented at a more perpendicular angle to the oncoming airflow and thus develops more lift. The rearward moving wing is presented at a more acute angle to the airflow and thus develops less lift. The result is that when a swept wing aircraft yaws it will experience a sharp rolling motion in the direction of the trailing wing.

Conversely a roll creates a yaw. As the aircraft sideslips during the roll the aircraft is induced into a yaw by the air pressure now acting on the fin. Because the aircraft is banked this yawing motion causes the aircraft nose to drop. At this point all three stabilities are being challenged. If left un-corrected the aircraft can enter a spiral dive. The area of the fin and its distance from the aircraft's centre of gravity has a bearing on the severity of this.

Stick Free Stability

We looked at this earlier in this chapter. When an aircraft is disturbed and moves around one or more of its axes we could assume that the pilot is holding the aircraft control column firmly and has his/her feet firmly planted on the rudder pedals. The stability of the aircraft in that situation is described as being 'stick fixed'. Now imagine that the pilot's hands and feet are clear of the controls. This is 'stick free'. If the aircraft is disturbed and moves, the hinged controls will experience inertia or a reluctance to follow the aircraft. This can result in the control 'trailing' and creating a further disturbing moment. For example if the aircraft nose were to pitch down and the elevator trailed, the aerodynamic force on the elevator would create a bigger nose down movement. Most large commercial aircraft today have power-operated controls and these will be firmly centred when the flight deck controls are centred. The problem occurs with manually operated controls. To reduce the inertial problems leading to control surface trailing, manually operated controls are mass balanced. You can say though, that in general terms aircraft that are being flown 'stick free' suffer a reduction in static stability.

Affect of Altitude

At altitude the tail plane is less effective in providing a restoring moment for a given true airspeed than at lower altitude. This reduces the longitudinal stability of the aircraft.

Dynamic Stability

Up to this point we have been examining the static stability of an aircraft. This is solely concerned with the ability of the aircraft to restore to its original attitude following a disturbance around one or more of its axes. Dynamic stability is concerned with the degree to which the aircraft may overshoot the original attitude and oscillate around it before finally stabilising. Imagine a pendulum being disturbed from its rest position. It is easy to see how it continually overshoots the original rest position in an attempt to stabilise again at rest. It may be statically very stable in trying to restore the original rest position but it has close to neutral dynamic stability in that it appears to continually oscillate about it.

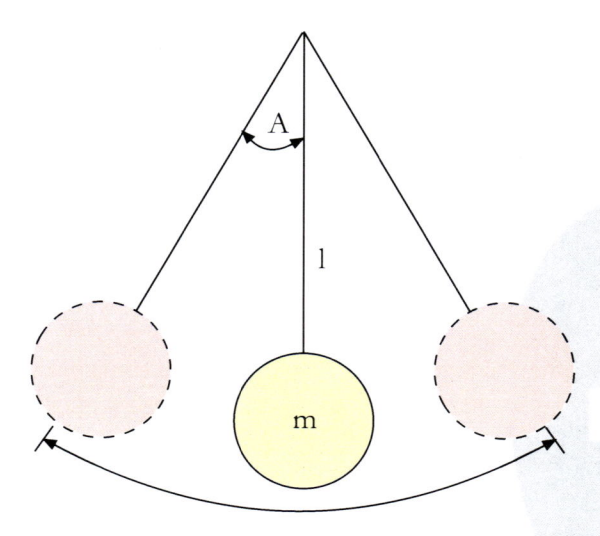

Figure 4.24 - Simple Pendulum

Imagine that the aircraft pitches down. The tail plane exerts a restoring moment that causes the aircraft to pitch up with sufficient momentum to push it back through the original longitudinally level trimmed position so that the nose is now pitched up. The tail plane now pushes the nose down again and once more we find that the aircraft overshoots back into a pitch down attitude. We could go on like this forever unless we can introduce a force to damp out the inertial effect. In case you have doubts about what is meant by 'inertia' I will give you a brief definition.

Inertia is the tendency of a body to remain at rest or if moving, to continue its motion in a straight line. The more mass a body has, the greater its inertia and the greater the force needed to overcome it to either accelerate it and give it momentum, or to decelerate it and destroy its momentum.

Momentum is a combined effect of inertia and velocity, or in other words, mass and velocity. The greater the mass and its velocity, the greater will be its momentum and thus the greater its inertia. While on the subject we need to look at another bit of physics.

We should examine a few more terms that we are going to encounter. If you consider a simple to and fro or oscillating movement that repeated itself in equal time intervals then you can describe that motion as being *Periodic*. The *Periodic Time* is the time taken for one complete to and fro movement, which we call a *Cycle*. The number of cycles occurring in one second is called the *Frequency* and this is measured in *Hertz (Hz)*, one Hz being equivalent to one cycle per second. The maximum displacement of a body from its middle or rest position is called the *Amplitude*.

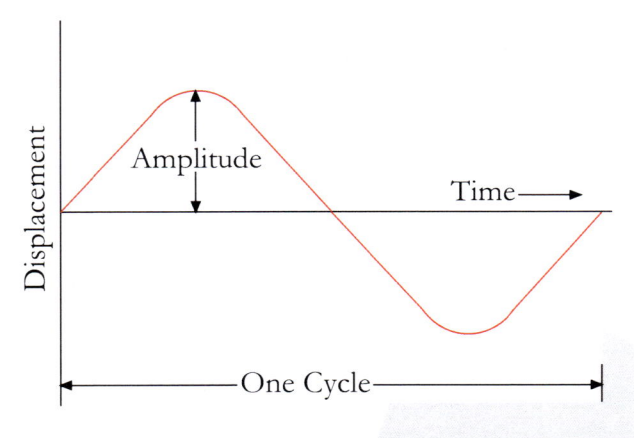

Frequency = Number of Cycles per Second

Period = Time to Complete One Cycle

Figure 4.25 - Periodic Motion

If the body always accelerates towards one fixed point in its path, like the mass of a swinging pendulum does towards its rest position, and the acceleration is proportional to the amplitude, the periodic motion is described as being *Simple Harmonic Motion*.

In examining dynamic stability we would hope the motion is not going to remain periodic. We would like to see the amplitude and periodic time decreasing as quickly as is possible as the aircraft recovers its original attitude. We need to damp the oscillating movement. Our success in doing this contributes to positive dynamic stability. Our failure to do it leads to negative dynamic stability. The thing to keep in mind is that an aircraft just like the pendulum can have very good static stability but very poor dynamic stability. It may quickly move back towards its original attitude after a disturbance but then swing around that position in its attempts to stabilise.

There are four classifications for this kind of stability. If the aircraft were to regain its original attitude with absolutely no overshoot or oscillation we would describe the dynamic stability as being positive as it is fully damped or '*Deadbeat*'. This is a pretty rare feat. The aircraft is '*Dynamically Stable*'.

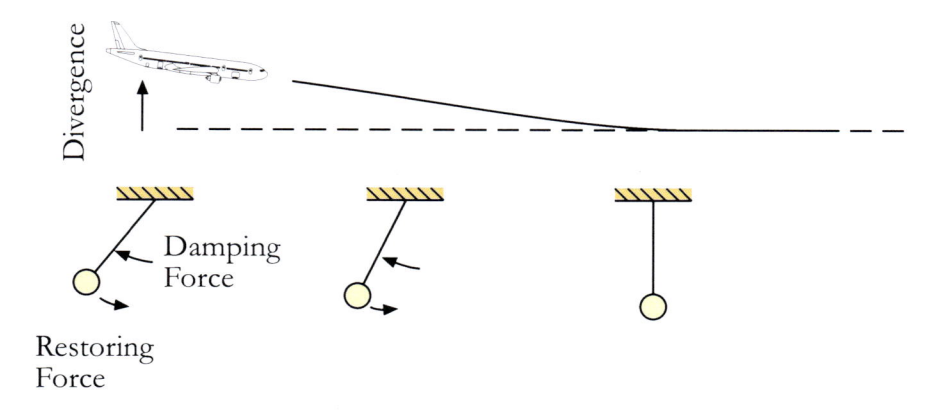

Figure 4.26 - Fully Damped Positive Stability (Dead Beat)

If the aircraft were to conduct say just two cycles before stabilising and the amplitude of the second oscillation was a lot smaller than the first we would describe this as being '*Short Period Damped*'. The aircraft would be considered to be '*Dynamically Stable*'.

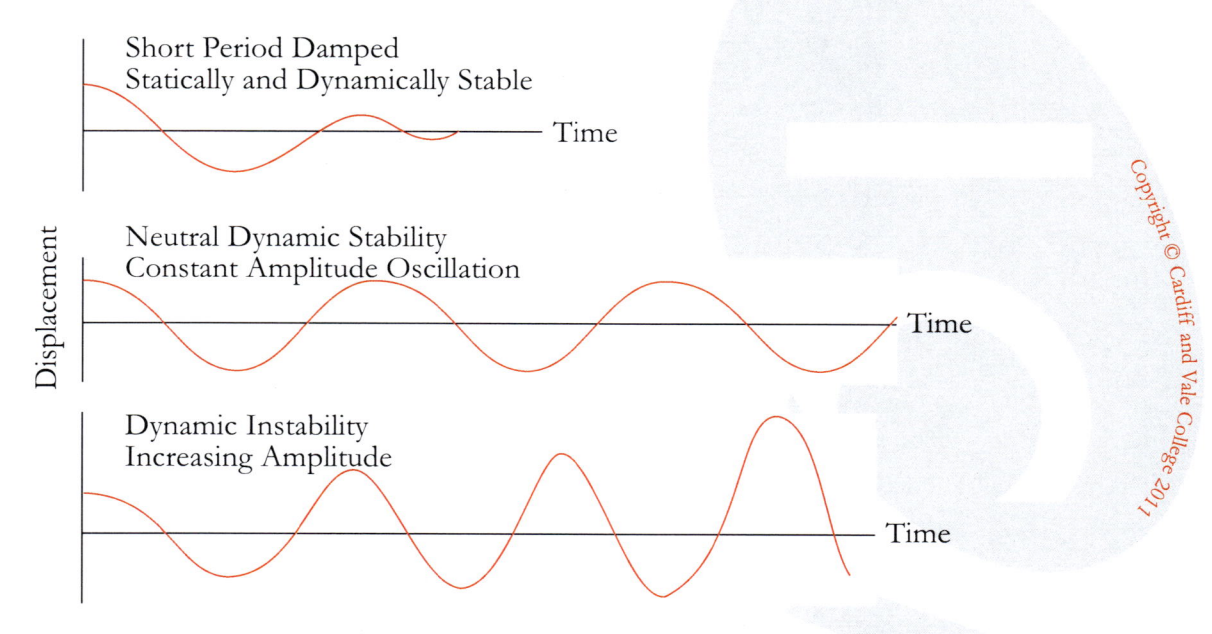

Figure 4.27 - Three Types of Motion

Now consider that the disturbed aircraft keeps on overshooting at constant amplitude from the original attitude line and continues to do this with no sign that the amplitude or frequency are changing. This is described as '*Neutral Dynamic Stability*'. It is periodic motion.

Finally, consider the situation where the aircraft overshoots and continues to do so with increasing amplitude and frequency. This is '*Dynamic Instability*'.

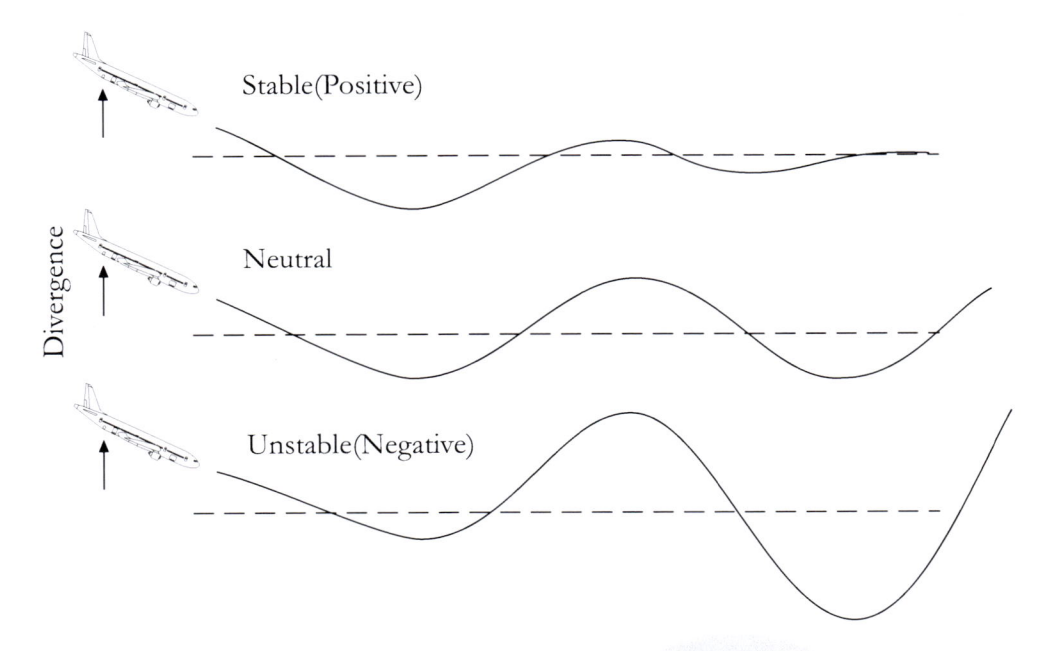

Figure 4.28 - Degrees of Dynamic Stability

Longitudinal Dynamic Stability

You will be aware by now that a disturbance to the aircraft's flight path that challenges its longitudinal stability will be an un-commanded pitch up or down. Let's see what happens when for example the aircraft pitches up. The vertical movement of the tail plane downwards will alter the direction of the relative airflow meeting it and this will increase the tail plane angle of attack to a greater value than if the tail had no vertical movement. So, while the tail is actually moving down there will be an increasing force working to resist the movement. This is a damping effect.

At the time this is occurring the aircraft's wings will be rotating around the centre of gravity position. Two things occur because of this. First the angle of attack of the wings will increase and this will increase the lift through the centre of pressure to strengthen the lift/weight couple that tends to push the aircraft nose down. Second while the actual pitching movement is occurring the trailing edge of the wing will have a downward motion whilst the leading edge will have an upward movement. This creates forces that act against the relative movements and will damp the pitching motion. The sum of these effects is that as the pitching movement is occurring there are forces resisting it that are in proportion to the rate and magnitude of the motion.

If the restoring pitching moment was just due to the increased angle of attack of the tail plane in pitch then the aircraft would merely oscillate around its original attitude and it would be neutrally stable. It is the damping forces that are created during the actual pitching movement that make it positively stable and produce the short period damping.

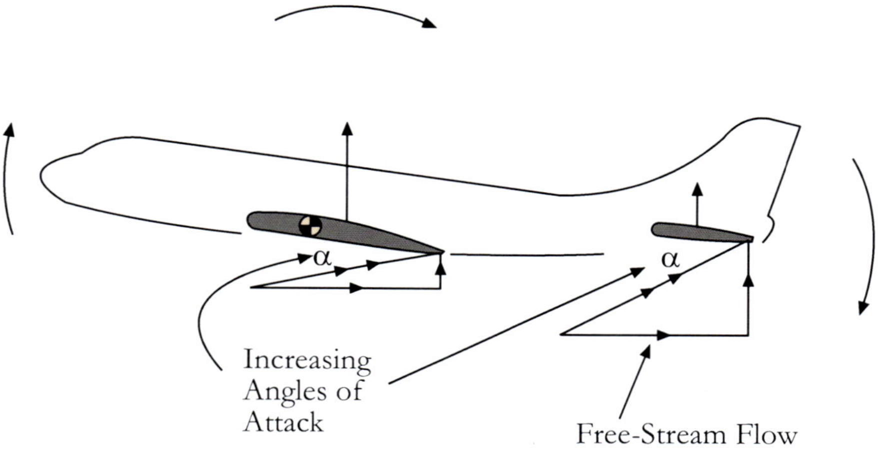

Figure 4.29 - Damping Action During Pitch

A disturbance leading to a pitch down movement produces similar damping effects. As the tail plane moves vertically upwards the relative airflow direction changes to reduce the angle of attack and thus the restoring force and will then produce a negative angle of attack if overshoot occurs to damp the overshoot. The rotation of the wing will reduce its angle of attack and weaken the lift/weight couple to give a nose up tendency. The opposite movements of the wing leading and trailing edges will also produce forces to resist the overshoot.

Imagine that I contrive to put a friction brake in the pivot of a pendulum with a rigid arm that acted to resist the periodic swing. I give the bob weight a push and the brake resists me. As the weight swings back towards rest, the brake resists this motion as well. The pendulum merely swings back to the vertical rest position and stops. It is both statically and dynamically stable. Useless of course if it were in your clock!

The pitch damping effects that we have examined should typically stabilise an aircraft in no more than one or at the most two cycles.

Effect of Altitude

Because the air density is relatively lower at altitude the aircraft has to fly faster to maintain the required lift. The increased free air stream velocity will influence the change in direction of the relative airflow onto the tail plane during pitching movements. This results in a smaller increase to the tail plane angle of attack during the actual movement and its damping action is reduced. So, longitudinal dynamic stability is reduced with altitude.

Phugoid (Porpoising)

A strange name but it refers to a second form of longitudinal dynamic instability that has a much longer period of oscillation. If you have ever flown paper aeroplanes you will have certainly seen it. If the aircraft is disturbed and pitches down as a result, you may recall from an earlier chapter that a component of the aircraft's weight would then act along the glide path. This has the effect of increasing the aircraft's speed. This causes the aircraft to accelerate on a descending sloped flight path. As the speed increases, so does the lift. This reduces the slope of the descent and eventually causes the aircraft to climb. Now the component of aircraft weight is acting against the climb and the aircraft begins to slow down again. If you haven't already guessed, it will suffer reduced lift and down it goes again! Unless the motion can be damped the aircraft will continue to oscillate between losing height and gaining height in a porpoise-like movement. The amplitude might remain constant and the stability neutral or, the amplitude may increase with increasing negative stability. The periodic time for this kind of oscillation is quite long, taking a minute or two to complete a cycle on a full size aircraft. The divergence can be startling at around plus and minus five hundred feet from the original flight path. Not a time to order lunch!

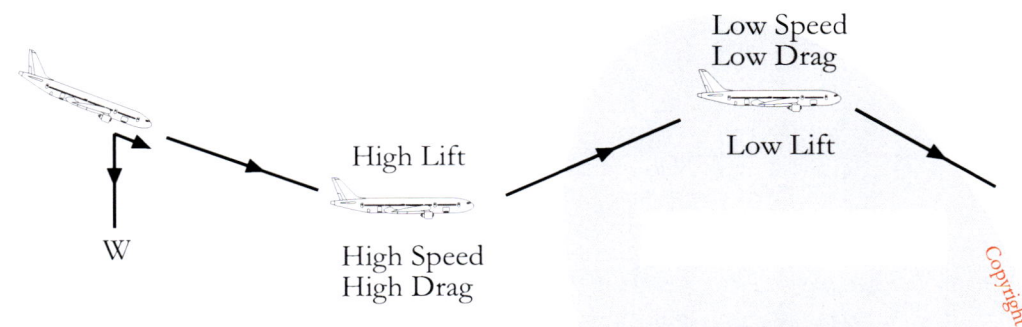

Figure 4.30 - Longitudinal Oscillatory Motion – The 'Phugoid'

The damping action that comes into play in this kind of motion is simply the change in drag on the aircraft as it oscillates. As speed increases in the descent drag increase up to a maximum at the bottom of the descent. As speed decrease to a minimum up to the top of the climb, drag decreases. In each case, the drag is resisting the changes in airspeed and altitude and is thus damping the oscillatory motion.

Lateral Dynamic Stability

We have already encountered the damping action that resists an aircraft rolling. The opposite vertical motions of the wings alters the angles of attack to increase lift on the down-going wing and decrease it on the up-going wing. Remember that this damping only occurs while the roll motion is in progress. The damping effect is proportional to the rate of the roll motion. As it is the vertical movement of the wings that creates this you may see that aircraft with short wingspans such as those with low aspect ratios like delta wings will not have as good a roll damping performance as higher aspect ratio wings.

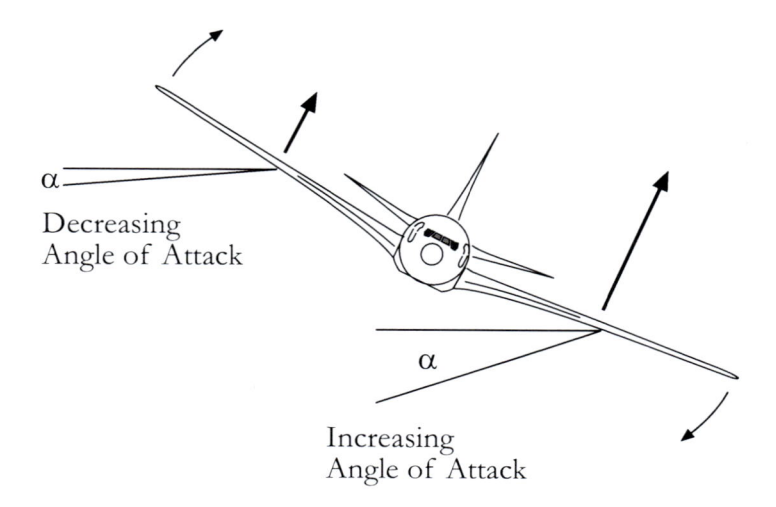

Figure 4.31 - Damping Action in Roll

With lateral stability we do have a couple of complications. You may recall that a roll motion creates a yawing motion and, a yaw creates a roll. We will examine the effects of this.

Dutch Roll

This is an oscillatory rolling and yawing motion that particularly affects swept wing aircraft. If the aircraft is disturbed and yaws, the fin will create a restoring moment. The aircraft will move back towards the original flight path but overshoots it by yawing in the opposite direction. This is in itself an oscillatory motion. The complication is that on the first yaw the aircraft also begins to roll in the direction of the yaw because of the increase in lift on the forward moving wing. The increase in drag that will also be experienced on the forward moving wing will create a force that attempts to make the aircraft yaw in the opposite direction. The initial rolling motion will reach a maximum coincident with the maximum rate of yaw. Unfortunately the sideslip tendency in roll reduces the yaw damping whilst the dihedral effect and the swept wing effect combine to push the aircraft back into an overshoot roll and yaw in the other direction helped by the drag on the wing on that side. The process then reverses and develops into a short period oscillatory rolling and yawing motion that is very unpleasant and reminiscent of the effects of a more than enthusiastic visit to the pub! Or so I have been reliably informed! This type of instability worsens as altitude increases.

Over-shoot on Yaw Corrections Create:

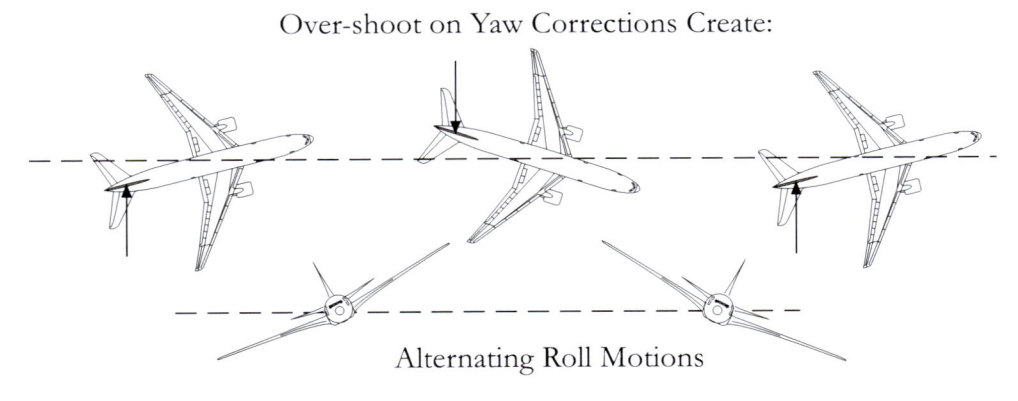

Alternating Roll Motions

Figure 4.32 - Oscillatory Roll and Yaw Motions of 'Dutch Roll'

The motion is fundamentally due to high lateral static stability or dihedral effect. This is why it will affect swept wing aircraft in particular. Although yaw damping actions do occur during roll and yaw they may be too weak to prevent the onset of the oscillatory motion. The damping can be improved by reducing the dihedral or introducing anhedral. Unfortunately reducing the lateral stability increases the influence of the fin and this makes the aircraft much more susceptible to yawing during a roll. Because the lateral stability is reduced the yaw can overcome it and the aircraft can enter a steep spiral dive.

This tendency is called spiral instability and designers do accept a small amount of it rather than suffer the more unpleasant Dutch Roll.

Some modern aircraft have automatic dual channel yaw damper systems fitted that use the auto-flight system to apply rapid control corrections to counter the initial yawing motions that invariably start imperceptibly. In this way Dutch Roll is prevented from initiating. Some aircraft are not permitted to fly with the automatic yaw damper systems inoperative.

As the yaw damper system is an active stabilising system I will briefly describe it for you. The system incorporates a rate gyro that is sensitive to the aircraft's motion in yaw. The coupling unit that contains the gyro can discriminate between controlled yaw demands and those un-controlled yaw motions that would lead to Dutch Roll. When the gyro senses the appropriate yaw movement a transducer converts it to an electrical signal that is then amplified and sent to a yaw damper actuator. This actuator signals the rudder actuator hydraulic jack to move the required amount to deflect the rudder to counter the yaw. A position transducer in the actuating jack sends a feedback signal to the rate gyro that cancels its signal. As the aircraft is restoring to its original heading the rate gyro senses this and sends out a further signal to progressively centre the rudder again. The system reaction is such that the rudder returns to its central position at a rate that prevents it from causing an overshoot into opposite yaw. The yaw damper system is extremely fast acting and can deal with very short cycle movements.

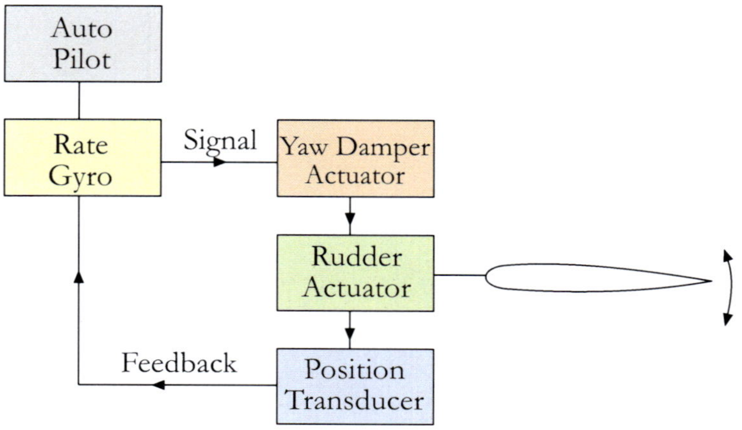

Figure 4.33 - Yaw Damper

Spiral Stability

We already know that the lateral stability of an aircraft depends on the restoring forces that will recover it to wings level when it is disturbed into a roll. We also have read that when the aircraft is rolling it tends to yaw into the direction of the lower wing. The problem is, that during the yaw the outer or upper wing develops more lift and this increases the rolling motion to a point where the lateral restoring forces may be overcome. If this occurs the aircraft angle of bank increases. A component of lift will now make the aircraft sideslip. The sideslip creates a side force on the fin and this in turn increases the yawing motion and the aircraft enters a descending spiral turn that can become increasingly steeper. The lateral and directional stabilities are interacting to create this situation. If this occurs we can say that the aircraft has spiral instability. Reducing the area of the fin and thus the directional stability will reduce the rate of the yaw into the sideslip thus reducing the lift on the outer wing in the turn. This improves the spiral stability. Conversely increasing the dihedral will increase the lateral restoring moment so that it cannot be overcome but this will make the aircraft more prone to Dutch Roll.

If the directional stability is higher than the lateral stability then spiral instability may occur. If the lateral stability is higher than the directional stability then Dutch Roll may occur. A compromise in favour of avoiding Dutch Roll is usually made.

The Spin

We have already examined the basic reasons underlying this motion. If an aircraft is disturbed in roll whilst near the stall and if the aircraft is prone to tip stall the lower wing will stall and drop. A yaw will immediately follow this, the motion being strengthened by the drag on the stalled wing. The aircraft will enter a spiral dive that is a combination of roll, sideslip and yaw. The rolling motion rather than the yaw predicts the steepness of the spin because the faster and further the lower wing descends, the greater its angle of attack and the deeper into the stall the lower wing goes. This leaves the upper wing with a reduced angle of attack but still producing lift that will increase the angle of bank. If the yawing motion is more predominant than the roll then the spiral dive will be flatter and less steep. The installation of mass balance weights in the nose and tail of the aircraft will flatten the spin because the turning motion of the aircraft will create centrifugal forces that will tend to move it towards the plane of rotation.

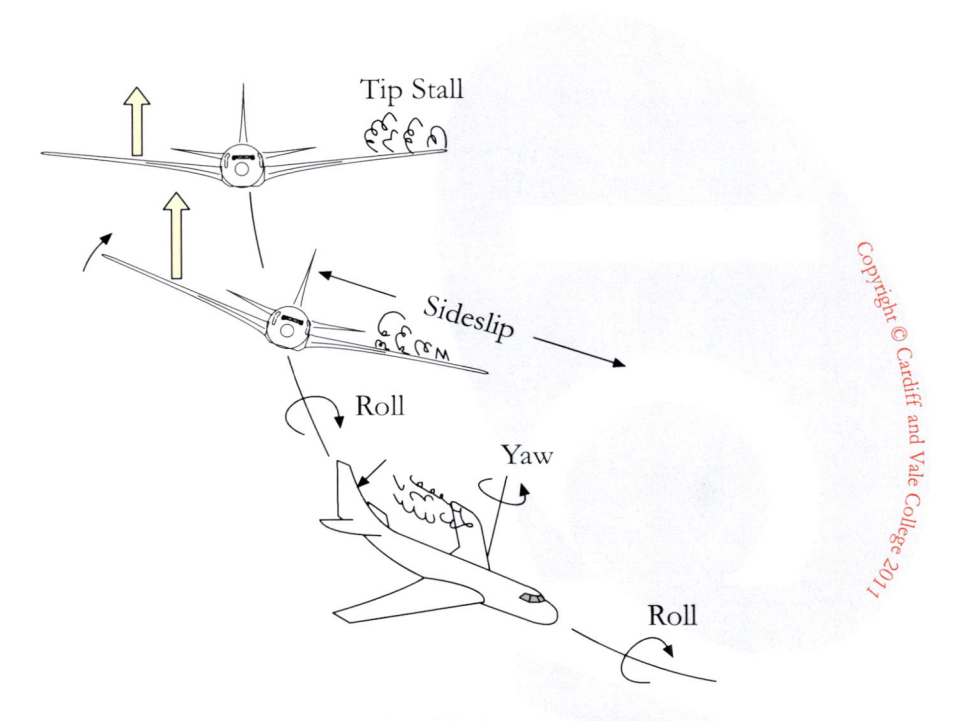

Figure 4.34 - Yaw and Roll Motions in a Spin

The yawing motion will allow the fin to create a **damping moment** that could balance the rolling moment and once this occurs the yaw rate steadies and the aircraft settles into a steady spin. Recovery from the spin can only be accomplished if the lower wing is brought out of its stalled condition. To do this the yaw must be stopped by the application of opposite rudder. The elevators are then used to pull the aircraft out of the dive.

Directional Dynamic Stability

We have already examined many of the facets of directional stability due to its strong associations with lateral stability. I must get to basics however and say that it is the keel surface to the rear of the aircraft centre of gravity including the fin that gives directional stability. The area of the fin and its distance from the centre of gravity are of course of paramount importance in influencing stability around the normal axis. It is important that the side fuselage centre of pressure is positioned aft of the aircraft centre of gravity. The area of the fin will determine this position. The fin has a symmetrical aerofoil section and when a yaw occurs in either direction the relative airflow direction onto the fin creates an angle of attack that in combination with the remainder of the keel surface aft of the aircraft centre of gravity resists the yawing motion and returns the aircraft to its original heading. Yaw in either direction is thus damped to give dynamic stability.

Stability of Propeller Driven Aircraft

We should examine a few problems associated with propeller driven aircraft and in particular single engine propeller driven aircraft. The slipstream from the propeller is spiral in nature and will strike one side of the fin to create an angle of attack that produces a side force that will induce a yaw. Various design features may be embodied to counteract this and would include for example off-setting the fin.

A further problem in relation to single engine propeller driven aircraft occurs during the take-off run. The reaction to the engine torque exerts a turning moment on the aircraft that is opposite to the rotation of the propeller. This exerts more pressure on one main wheel and causes the aircraft to veer to the side of the runway. A propeller rotating clockwise viewed from the rear exerts pressure on the port main wheel causing the aircraft to veer to port on take-off. This only occurs whist the aircraft weight is on the ground. If you add this effect to the slipstream effect we discussed earlier, the pressure on the fin will also be causing a yaw tendency to port.

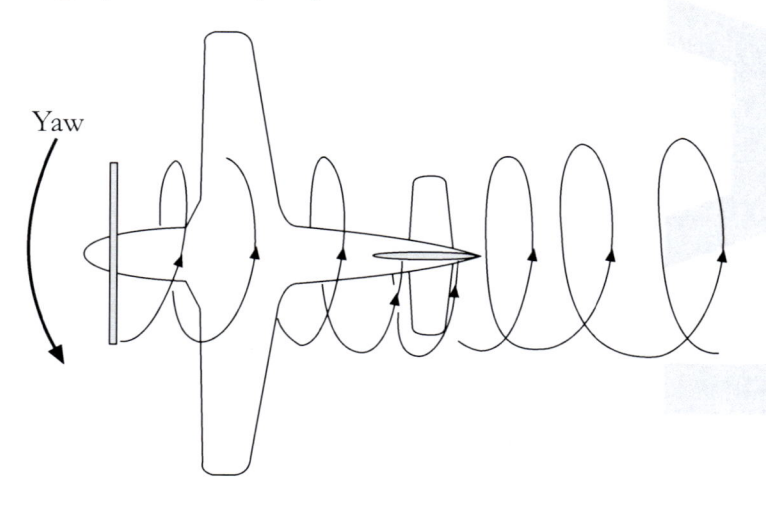

Figure 4.35 - Effect of Propeller Slipstream

Lastly, propeller driven aircraft experience a gyroscopic effect called 'precession' when the direction of the aircraft is changed. Simply, when you apply a force to change the plane of rotation of a spinning object like a propeller, the force transfers ninety degrees around the circumference of the propeller in the direction of rotation and the propeller then tries to veer in the direction indicated by the force. From our example above, if the aircraft nose were lifted during flight the aircraft would tend to yaw to starboard. It is interesting to note that if our example single engine aircraft had a tail wheel the aircraft nose would lower as the tail lifted on the take-off run and the precession force would create yet a further tendency to veer to port. Some twin-engine aircraft are designed so that the port and starboard engine propellers rotate in opposite directions and this cancels out the effects.

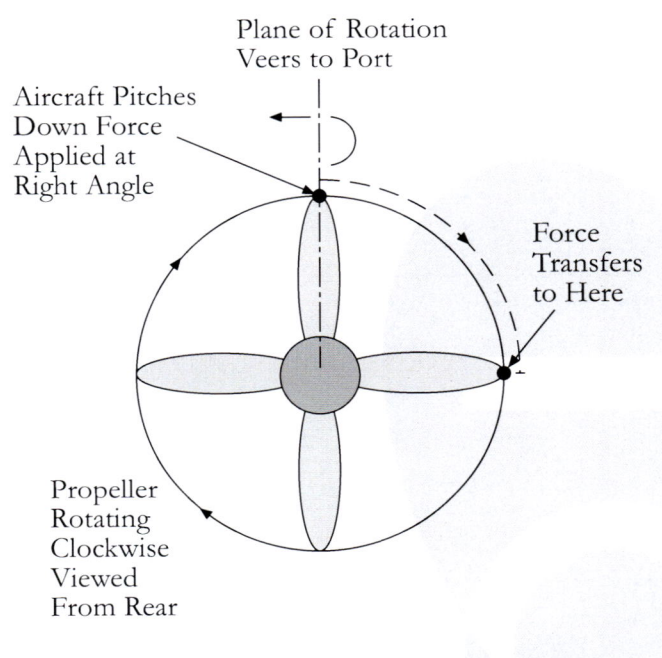

Figure 4.36 - Gyroscopic Effect of a Propeller

Stability Design Features

As a final reminder make sure that you can identify the **main design** features that produce the restoring forces required for stability.

Stability	Axis	Design Feature
Longitudinal(Pitch)	Lateral	Tail-plane Angle of Attack
Lateral(Roll)	Longitudinal	Dihedral and Sweepback
Directional(Yaw)	Normal	Keel Surface aft of Aircraft C of G

Table 4.2 – Stability Design Features

CHAPTER FOUR
FLIGHT STABILITY & DYNAMICS

Unstable Aircraft

It is not a good thing to have an aircraft that possesses too great a stability. The aircraft would not display a satisfactory response to control inputs for manoeuvres. It would fight you. A degree of instability can be tolerated in return for an aircraft that is more responsive. Aircraft that are used for basic flying training may have built in instability in order that students can learn to fly the aircraft rather than it fly them. This way they can become experienced in the bad habits of aircraft with low stability.

It is becoming more common to see aircraft that are designed to be deliberately unstable. This is predominantly the case with recent military aircraft but is creeping into the civil aircraft arena. There are advantages to be gained and you already know of one, control response. If you move the centre of gravity back to the neutral point we also know that this produces neutral stability and removes the tail plane correcting forces so we have no trim drag. If the centre of gravity is moved aft of the neutral point the tail plane is then required to produce lift. This means the tail plane and the wings are both producing lift and we get a much better overall lift/drag ratio for this. This means that the centre of gravity is either very close to the centre of pressure position or moves aft of it. The payback for suffering the resultant instability is vastly improved performance in terms of lift/drag ratio and control agility. Where is the catch?

Unstable aircraft require hard work on the part of the pilot in order that the aircraft flies safely. This reaches the point where the pilot cannot fly the aircraft because his/her response times are not fast enough. For example, if an aircraft is oscillating around in pitch, the pilot may get the sensation of a pitch down and then decide to apply up elevator to correct it. By the time this has been thought about and the elevator control input has been applied the aircraft is pitching up anyway and the pilot has just made the pitch up movement greater. In attempting to continually correct oscillations and getting out of step the pilot is inducing even higher amplitude oscillation. This has a name, it is called **Pilot Induced Oscillation (PIO)** and it can quickly put an aircraft into an irretrievable attitude.

There has to be a solution and there is.

We now have extremely reliable electronic flight control systems that in conjunction with hydraulic power operated controls are able to control aircraft that would otherwise be uncontrollable. For safety reasons this requires that these electronic control systems are duplicated and more commonly triplicated. Many unstable aircraft are now being flown quite satisfactorily by 'fly by wire' systems that could not be flown by a human pilot. The next time you see a well-known European aircraft type conducting a display you may notice how it is being flown rock steady even when it is right on the operational limits of pitch, roll and yaw. Impressive, but no human can hold an aircraft safely at those limits. Five independent computers are doing it! I will give you a brief introduction to a simple auto-flight system as they are increasingly being used to introduce active control and stability in addition to the yaw dampers and the gust alleviation systems described earlier.

Basic Autopilot

The autopilot system relieves the flight deck crew of the work associated with the continual control of the aircraft and the maintenance of altitude, airspeed and navigation headings. The system is also sensitive to stability and will incorporate damping systems that detect inertial forces leading to divergence or oscillation around the three axes of rotation.

Most autopilot systems are of the two-axis configuration, controlling the elevator and aileron/spoiler movements. The rudder will only have an automatic yaw damper system because the ailerons can induce a change of direction, as you know a roll will create a yaw.

The auto-flight system incorporates gyro displacement and rate of displacement sensors, computers, servo actuators and a control unit to permit selection from a menu of auto-flight functions. The control unit also gives the pilot the ability to input trimming commands or to initiate manoeuvres. It also receives radio navigation signals, pre-selected heading information from the horizontal situation indicator setting, altitude **selections** and even airspeed selections in conjunction with the engine **throttle** computer. More recent systems are linked directly to the pilot's flight **deck controls and receive** signals related to their movement that are computed and resolved into control surface movements. This is the 'fly-by-wire' concept. The computers are programmed with the aircraft aerodynamic and structural limits. If a suicidal pilot pulled fully back on the control column and also applied full aileron the computers would merely put the aircraft up to the maximum safe angle of attack (Alpha Floor) and the maximum safe angle of bank at the maximum safe rate of roll for that condition – all the time maintaining stability. Impressive!

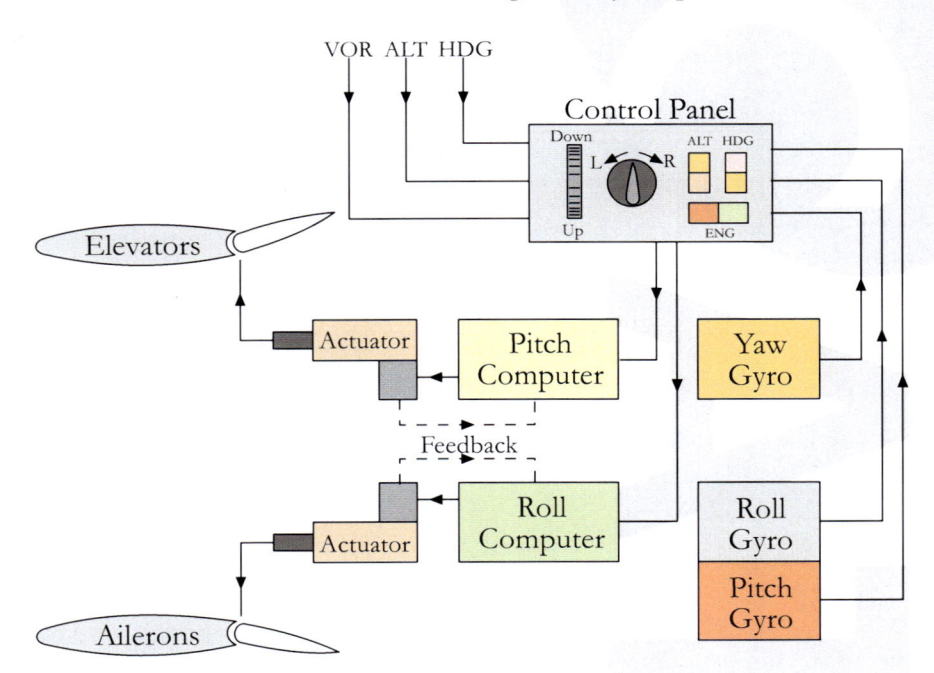

Figure 4.37 - Two-Axis Auto-flight System

CHAPTER FOUR
FLIGHT STABILITY & DYNAMICS

The flight management system now incorporates many functions previously conducted by the pilot. Selection of the ambient corrected maximum engine thrust for take-off, cruise thrust, climb thrust and 'go-around' engine and flight functions. The system is connected to the Automatic Direction Finder (ADF), The VHF omni-directional range navigation beacons (VOR), the long range navigation transmissions (VLF/OMEGA), the inertial navigation system (INS), the instrument landing system (ILS), the Doppler navigation system. But it will not serve you a cup of coffee – yet.

The basic principle is simple. Movement of the aircraft is detected by displacement and rate gyros in terms of direction, magnitude and rate. These are similar to the gyros used in the horizontal situation indicator (HSI) for direction and the attitude situation indicator (ASI) or artificial horizon for roll and pitch. Sensors installed in the gyros are transducers that convert the gyro precessions into electrical signals and send them to the appropriate computers. The computers process the signals and condition them into electrical signals that are sent to the appropriate flying control hydraulic actuators. These actuators contain position sensors that relay 'feed-back' signals to the computers telling them that the commanded movements have been accomplished. All the time the gyros are monitoring the restoring movements and the inertial forces resulting from rates of movement and continually supply the computers with this information to maintain the aircraft's static and dynamic stability.

This is intended to be just a very general familiarisation with these systems as it is doubtful that any in depth questioning will be encountered on this topic in the module examination as it would be crossing over into the avionic system module areas.

Conclusion

Do try the questions that follow. As a personal postscript only I have dedicated this publication to Jill, a four-legged friend, who was my constant companion while writing this and is sadly no longer here. I wish you every success in your future examination.

Revision

Flight Stability & Dynamics

Questions

1. **The static stability of an aircraft is its:**

    a. balance around its centre of gravity

    b. manoeuvrability

    c. ability to return to its trimmed position after a disturbance

2. **If an aircraft returns towards its trimmed position following a disturbance but then oscillates around it, the aircraft is described as having:**

    a. static instability and dynamic stability

    b. static stability and dynamic instability

    c. static and dynamic instability

3. **An aircraft is described as having lateral stability around the:**

    a. longitudinal axis

    b. lateral axis

    c. normal axis

4. **An aircraft pitches around the:**

    a. lateral axis

    b. longitudinal axis

    c. normal axis

5. Dihedral improves:

   a. longitudinal stability

   b. lateral stability

   c. directional stability

6. During a sideslip, the restoring moment is primarily produced by the:

   a. tail plane

   b. keel surface

   c. dihedral

7. Longitudinal stability is provided by the:

   a. fin

   b. dihedral

   c. tail plane

8. The 'Phugoid' or porpoising oscillation is an instability in:

   a. pitch

   b. yaw

   c. roll

9. Dutch Roll is an oscillatory motion that is an interaction between:

   a. longitudinal and directional stability

   b. lateral and directional stability

   c. longitudinal and lateral stability

10. **Dihedral produces a restoring force during a:**

    a. roll

    b. sideslip

    c. yaw

11. **A high mounted wing configuration displays:**

    a. improved longitudinal stability during pitching

    b. decreased directional stability during a sideslip

    c. increased lateral stability during a sideslip

12. **When a conventional aircraft pitches up the increase in lift will tend to:**

    a. push the nose down

    b. push the nose up

    c. not effect the pitch attitude

13. **The fin will provide:**

    a. directional stability around the normal axis

    b. longitudinal stability around the normal axis

    c. directional stability around the lateral axis

14. **Differential ailerons will:**

    a. increase profile drag on the down-going aileron

    b. increase induced drag on the up-going aileron

    c. equalise the total drag of both ailerons

15. When the centre of gravity is moved rearwards on a conventional aircraft the:

    a. static stability improves

    b. static stability decreases

    c. dynamic stability improves

16. If the downwash increases the effect on the tail plane will:

    a. increase longitudinal static stability

    b. decrease longitudinal static stability

    c. increase longitudinal dynamic stability

17. When roll is predominant over yaw in a spin the aircraft will:

    a. enter a steeper dive

    b. flatten out of the dive

    c. spin slower

18. If an aircraft yaws to the left it will:

    a. roll to the right

    b. roll to the left

    c. remain wings level

19. If the thrust is increased on a conventional aircraft it will tend to:

    a. pitch down

    b. remain straight and level

    c. pitch up

CHAPTER FOUR
FLIGHT STABILITY & DYNAMICS

20. The 'pendulum effect' on a high wing aircraft will:

    a. increase its lateral static stability

    b. decrease its lateral static stability

    c. increase its lateral dynamic stability

21. If an aircraft is disturbed and oscillates at constant amplitude and frequency it is said to have:

    a. dynamic instability

    b. dynamic stability

    c. neutral dynamic stability

22. When an aircraft recovers from an un-commanded roll it is said to have:

    a. directional stability

    b. longitudinal stability

    c. lateral stability

23. The angle of attack on the down-going wing during a roll will:

    a. decrease

    b. increase

    c. not alter

24. Wing sweep-back:

    a. increases lateral static stability

    b. decreases lateral static stability

    c. has no effect on stability

25. Lateral static stability can be increased by:

    a. increasing anhedral

    b. increasing dihedral

    c. reducing wing sweep angle

26. A positive longitudinal dihedral requires that the:

    a. wings and tail plane have similar angles of incidence

    b. wings have a smaller angle of incidence than the tail plane

    c. tail plane has a smaller angle of incidence than the wings

27. Directional stability is given by the:

    a. rudder only

    b. the fin only

    c. the fin and keel surface

28. Longitudinal stability is achieved when:

    a. the centre of gravity is to the rear of the centre of pressure

    b. the centre of pressure is to the rear of the centre of gravity

    c. the centre of pressure is coincident with the centre of gravity

29. If the aircraft centre of gravity is moved to its forward limit the controls will be:

    a. less responsive

    b. more responsive

    c. unaffected

30. If an aircraft is disturbed and continues to diverge away from its original position it is said to have:

    a. static instability

    b. neutral static stability

    c. stability

31. Dutch Roll is a combination of:

    a. pitch and yaw

    b. roll and pitch

    c. yaw and roll

32. The three axes of an aircraft pass through the:

    a. centre of pressure

    b. centre of gravity

    c. aerodynamic centre

33. Directional dynamic stability may be increased by the use of:

    a. lateral dihedral

    b. yaw dampers

    c. pitch dampers

34. Anhedral is used on some swept wing aircraft to:

    a. improve the lateral control response

    b. increase lateral stability

    c. increase spiral stability

35. The stability of an aircraft related to increasing altitude:

    a. increases

    b. remains constant

    c. decreases

36. The aerofoil section of an aircraft fin is:

    a. positively cambered

    b. negatively cambered

    c. symmetrical

37. The slipstream from a single engine propeller driven aircraft will:

    a. induce yaw

    b. will not induce yaw

    c. reduce the effectiveness of the fin

38. Compared to a low wing aircraft, a high wing aircraft has:

    a. poor lateral stability

    b. better lateral stability

    c. similar lateral stability

39. Compared to holding the controls in a fixed centralised position, stick free stability is:

    a. better

    b. worse

    c. similar

40. The lateral stability of the delta wing aircraft compared to high aspect ratio wings is:

   a. similar

   b. better

   c. worse

CHAPTER FOUR
FLIGHT STABILITY & DYNAMICS

Revision

Flight Stability & Dynamics

Answers

1.	C	21.	C
2.	B	22.	C
3.	A	23.	B
4.	A	24.	A
5.	B	25.	B
6.	C	26.	C
7.	C	27.	C
8.	A	28.	B
9.	B	29.	A
10.	B	30.	A
11.	C	31.	C
12.	A	32.	B
13.	A	33.	B
14.	C	34.	A
15.	B	35.	C
16.	B	36.	C
17.	A	37.	A
18.	B	38.	B
19.	C	39.	B
20.	A	40.	C

Glossary of Terms

Adverse Yaw

A tendency of an aircraft to yaw in a direction that is opposite to the direction of a controlled turn as a result of an imbalance in induced drag created by deflection of the ailerons

Aerodynamic Centre

A reference point on the chord-line of a wing, about which the pitching moment remains constant regardless of the angle of attack

Aerofoil

A body shaped so as to produce an aerodynamic reaction called lift that acts perpendicular to its direction of motion or line of flight

Aileron

The control surface used for lateral control

Airspeed

The speed of an aircraft relative to the air

Angle of Attack

The angle formed between the chord line of an aerofoil and the relative airflow direction

Angle of Incidence

The constructed angle formed between the longitudinal datum of the fuselage and the chord line of the wing

Anhedral

The downward slope of a wing relative to the horizontal plane designed to reduce lateral stability

Aspect Ratio

The ratio of the wing span to the average or mean chord. Can also be the ratio of the square of the wing span to the total wing planform area.

Axis

A line about which the aircraft rotates

Bank

The attitude adopted by an aircraft in a turn where the lift supplies a component towards the centre of the turn called centripetal force

Bernoulli's Theorem

States that the total energy contained in an incompressible flow of air through a venturi will remain constant

Boundary Layer

The layer of air flowing over a surface where shearing action takes place between the surface and the free stream air velocity that retards the flow as the surface is approached

Camber

The curvature of an aerofoil represented by a line drawn equidistant from the upper and lower surfaces

Centre of Gravity

The point through which the weight of a body acts whatever position the body may be in

Centre of Pressure

The position on the chord at which the resultant of all the lift forces acts

Centripetal Force

The continuous force that keeps a body travelling in a circular path

Chord

A straight line that joins the centre of curvature of the leading edge of an aerofoil section to the centre of curvature or apex of the trailing edge

Coefficient

A numerical constant that is used as a multiplier when calculating a variable quantity such as lift or drag

Couple

Two equal and opposite forces that act on a body that are not in the same straight line but whose moment about any point in the plane is always the same. A single force cannot balance a couple. It can only be balanced by an opposing couple of equal magnitude.

Density

Mass per unit volume

Dihedral (Lateral)

The upward slope of a wing relative to the horizontal plane designed to increase lateral stability

Dihedral (Longitudinal)

The angle between the chord of the tail plane and the chord of the main plane designed to give longitudinal stability

Directional Stability

The ability of an aircraft to regain its original position following a divergence in yaw around the normal axis. Stability is provided by the keel surface aft of the CG.

Downwash

The downward deflection of the airflow in the wake of a wing

Drag

The force that opposes the forward motion of a body through the air

Dutch Roll

A roll and yaw motion. Occurs when lateral stability is higher than directional stability particularly on swept back wing aircraft

Dynamic Stability

The ability of an aircraft to return to its original position after a disturbance without oscillating around that position

Elevator

The control surface used for longitudinal control

Endurance

The ability to remain in the air for the longest possible time

Equilibrium

The state of even balance where all opposing forces and turning moments neutralise each other

Equivalent Airspeed (EAS)

The airspeed calculated from the measured pressure difference between ambient and sea level ISA

Fences

Vanes fitted chord-wise across a swept back wing to check the span-wise boundary layer outflow

Fin

A vertical symmetrical aerofoil section structural member designed to provide directional stability. Alternative name is vertical stabiliser.

Fineness Ratio

The length of a streamlined shape divided by its maximum thickness

Flap

A lift augmentation surface designed to increase the coefficient of lift throughout the normal range of the angle of attack

Form Drag

Boundary layer normal pressure drag resulting from the adverse pressure gradient over the back of an aerofoil creating a higher-pressure region behind the low pressure region at the front. It forms a part of the profile drag.

Fowler Flap

A flap that moves rearwards initially increasing area and then downwards to increase the camber of a wing

Free Stream Air

The air that is undisturbed by the passage of a body through it

Frise Ailerons

A configuration where the up-going aileron projects a 'beak' into the airflow under the wing to create additional profile drag to counter adverse yaw

Horizontal Stabiliser

The tail plane. A structural member fitted to give longitudinal stability.

Horse Shoe Vortex

The shape derived when the wing tip trailing vortices are bridged in the aircraft wake by the wing bound vortex so forming the three sides of a closed figure

Indicated Airspeed (IAS)

The airspeed displayed on the air speed indicator that due to changes in air density will not always correspond with the true air speed or the equivalent lower than true airspeed at altitude.

Induced Drag

A type of drag that is generated as a by-product of lift and is inversely proportional to the square of airspeed. If airspeed doubles, induced drag reduces by a factor of four.

Interference Drag

A part of Profile Drag that is produced by the effect of the airframe component joints where changes in section create interference to the local boundary layer airflows. Wing and tail to fuselage etc.

International Standard Atmosphere (ISA)

A table of standard values for atmospheric conditions at different altitudes that is used internationally to establish and compare the performance of equipment that is reliant on air density

Inertia

The tendency of a body to remain at rest or if moving to continue moving in a straight line. The greater the mass of the body the greater its inertia will be.

Keel Surface

The area presented by the aircraft in side elevation. The area to the rear of the C of G gives directional stability.

Laminar Flow

The boundary layer flow where the streamlines remain separated and the varying speed sub-layers slide over each other without intermingling

Lateral Axis

A line running span-wise through the aircraft centre of gravity at right angles to the longitudinal and normal axes

Lateral Stability

The ability of the aircraft to return to its original lateral axis around the longitudinal axis called roll. Stability is provided by dihedral and sweep back.

Leading Edge

The front edge of an aerofoil or streamlined shape

Lift

An upward force whose line of action is at right angles to the relative airflow direction and acts on the Centre of Pressure. Lift = $C_L \times 1/2\rho v^2 \times S$

Lift Coefficient (C_L)

A measure of the lift effectiveness of an aerofoil that takes into account shape and angle of attack. Used as a multiplier to the lift calculation $1/2\rho v^2 S$

Lift/Drag Ratio

The ratio of total lift and total drag L/D

Longitudinal Axis

A line running from the aircraft nose to tail that passes through the centre of gravity at right angles to the normal and lateral axes

Longitudinal Stability

The ability of an aircraft to recover to its original position following a disturbance around the lateral axis that is called pitch. Stability is provided by the tail plane angle of attack.

Mach No.

The true air speed of an aircraft divided by the local speed of sound in air

Manoeuvre point

The aft centre of gravity position where any movement of the elevator would cause instability

Manoeuvre Margin

The distance from the aircraft centre of gravity to the Manoeuvre Point

Mean Chord

The average chord length of a wing

Moment

The moment of a force about a point is the product of the force and the perpendicular distance between its line of action and the point

Neutral Point

The aft position of the aircraft centre of gravity where the aircraft is on the point of becoming unstable. It is point where longitudinal stability becomes neutral.

Optimum Angle of Attack

The angle of attack where an aerofoil would produce the highest lift/drag ratio, usually 3° to 4°

Oscillation

To swing to and fro, like a pendulum

Phugoid

Otherwise known as 'porpoising'. An oscillatory diving and climbing long period motion

Pitch

A nose up or down movement when the longitudinal axis moves around the lateral axis

Pressure

Load divided by cross-sectional area. Nm^2 or $lb\,in^2$

Profile Drag

The drag associated with the shape of a body and its surface finish. Includes Form Drag, Skin Friction and Interference Drag. Combines the effects of boundary layer friction drag and surface friction. Profile drag is proportional to the square of airspeed.

Relative Airflow

The airflow in relation to the aircraft passing through it

Reynold's No.

The value where laminar flow becomes turbulent. Depends upon air density, air velocity, chord length and coefficient of air viscosity. $\rho v l/\mu$.

Roll

The movement of the lateral axis around the longitudinal axis

Rudder

The control surface used for directional control

Separation

Describes the action where the flow no longer follows the contour of an aerofoil but 'separates' from it to produce a turbulent re-circulating airflow

Sideslip

The sideways movement produced by a component of lift produced when an aircraft rolls

Skin Friction

The frictional force produced as a result of the shearing action between the layers of varying speed air flows in the boundary layer over the surface of an aircraft. It is a component of profile drag.

Short Period Pitching Oscillation

A heavily damped oscillation that incurs little change in aircraft height or speed and lasts for no more than one or two cycles

Sink Rate

Loss of altitude with time during a gliding descent for endurance

Slat

A leading edge device used to control the boundary layer and prevent leading edge separation. Increases the stalling angle of attack of an aerofoil.

Slender Delta

A type of delta wing designed to operate with separated flow over the wing designed to produce vortex lift

Slot

A device used to control the boundary layer and prevent flow separation. May be used as a leading edge device or with trailing edge flaps

Sonic

The condition where all the airflow over a body is travelling at the local speed of sound in air

Speed of Sound in Air

The speed at which sound waves, travel through air. The speed is proportional to the absolute temperature of the air only.

Span

The length of a wing taken at right angles to the longitudinal axis to the wing tip. Wingspan is the total straight-line distance across the aircraft from wing tip to opposite wing tip.

Spin

A spiral dive created when directional and lateral stabilities become unbalanced usually initiated by a wing tip stalling

Spiral Instability

Occurs when directional stability is much higher than lateral stability. A yaw produces a roll that produces a sideslip that produces a further yaw into a spiral dive.

Spoilers

Control surfaces consisting of a hinged flaps on the top wing surface that can be used to differentially spoil lift and support the ailerons in lateral control or collectively to act as speed brakes in the air or lift dumpers on the ground

Static Stability

The ability of an aircraft to recover to its original position following a disturbance around its axes

Stagnation

A region of stationary air usually occurring just below the leading edge of an aerofoil where the airflow divides to pass on either side

Stall

The point where the boundary layer separates from the surface resulting in a sharp loss of lift. Occurs around fifteen degrees angle of attack on an aerofoil.

Stalling Angle

The angle of attack of an aerofoil at which stall occurs

Starting Vortex

A horizontal vortex that is formed and left behind the aircraft whenever there is an increase in wing circulation such as on take-off and when pulling out of a dive. Wing bound vortex.

Static Pressure

The ambient atmospheric pressure

Streamlines

The lines traced out in smoke delineating smoothly flowing layers of air in laminar flow

Streamlined shape

A shape where the air flows around it in streamlines without separating and becoming turbulent

Subsonic

The condition where all the airflow over a body is travelling at a speed below the local speed of sound in air

Supersonic

The condition where all the airflow over a body is travelling at a speed above the local speed of sound in air

Superstall

The locked in stall that occurs when a tailplane that is mounted on the top of a fin is totally enclosed in the turbulent wake of stalled main planes rendering longitudinal control ineffective.

Tailplane

The structural component that is fitted to give longitudinal stability. Sometimes referred to as the horizontal stabiliser.

Taper Ratio

The ratio of the chord lengths of the wing root and the wing tip. Taper ratio zero refers to a pointed tip.

Thrust

The forward acting propelling force that opposes drag

Total Reaction

The resultant of the lift and drag forces acting on an aircraft

Trailing Edge

The rear edge of an aerofoil

Trailing Vortex

The vortex formed at a wing tip resulting from the airflow under the wing trying to curl over into the low-pressure region above the wing. This vortex is the primary cause of induced drag. Its intensity is inversely proportional to airspeed and proportional to aircraft weight.

Transition point

The point on the surface of an aerofoil where laminar airflow becomes turbulent. The point moves forward as airspeed increases.

Transonic

The condition where the airflow velocity over a body is part sub-sonic and part sonic or supersonic.

Trim Drag

The profile drag created by the tail plane.

Turbulent Flow

The situation where streamlines cannot maintain their separation and intermingle creating vortices and re-circulatory flow

Upwash

The upward deflection of the airflow at the leading edge of an aerofoil

Viscosity

A fluid's resistance to flow

Vortex

A spinning column of air with a low-pressure core

Wash-in

An increase in the angle of incidence of a wing from its root to its tip

Wash-out

A decrease in the angle of incidence of a wing from its root to its tip

Weight

The force that acts vertically downwards through the centre of gravity that is the product of the mass and the acceleration due to gravity.

Winglet

Aerofoil shaped device fitted at each wing tip to modify and raise the wing tip trailing vortices to reduce their influence on induced drag.

Wing Sweep Angle

The angle formed between the longitudinal axis and a line drawn along the wing at one quarter chord length.

Yaw

The movement of the longitudinal axis around the normal axis. Stabilised by the fin and controlled by the rudder.

Yaw dampers

An automatic system used to damp un-commanded yawing motions that would otherwise lead to Dutch Roll. Used as an alternative to reducing lateral stability on some swept wing aircraft.